Modeling Water Quality
in
Drinking Water Distribution Systems

Modeling Water Quality
in
Drinking Water Distribution Systems

Robert M. Clark
Walter M. Grayman

American Water Works Association
Dedicated to Safe Drinking Water

Denver ❖ 1998

Modeling Water Quality in Drinking Water Distribution Systems

Editor: Mindy Burke
Production Manager: Scott Nakauchi-Hawn
Cover Design: Karen Staab

Library of Congress Cataloging-in-Publication Data

Clark, Robert Maurice.
    Modeling water quality in drinking water distribution systems /
Robert M. Clark, Walter M. Grayman
        xv, 231 p. 17×25 cm.
        Includes bibliographical references and index.
        ISBN 0-89867-972-9
        1. Water quality--Computer simulation. 2. Water--Distribution-
    -Computer simulation.   I. Grayman, W. M.   II. Title.
    TD482.C58   1998
    628.1'61--dc21   98-36443
                                                        CIP

Printed in the United States of America

American Water Works Association
6666 West Quincy Avenue
Denver, CO 80235

ISBN 0-89867-972-9

Printed on recycled paper

Contents

List of Figures

List of Tables

Foreword

The Safe Drinking Water Act (SDWA) of 1974 and its amendments in 1986 mandated that the US Environmental Protection Agency (USEPA) establish maximum contaminant levels (MCLs) for a selected set of contaminants in drinking water. Establishing the MCLs involves considering health effects, available technology, monitoring feasibility, and risk assessment. A number of comprehensive rules, such as the Surface Water Treatment Rule (SWTR), also have been established under the SDWA. Amendments to the SDWA in 1996 emphasized small systems, source water protection, and the establishment of a state revolving loan fund (SRLF) to pay for needed investments in treatment and distribution systems.

Most SDWA regulations to date have focused on treating water, even though water quality can deteriorate once in the distribution system. For example, there are numerous ways bacteria can enter a drinking water distribution system. These avenues include open reservoirs; enclosed, unchlorinated reservoirs; new construction that disturbs the existing distribution system; main breaks (which will become an increasing problem as systems age); backpressure; dead ends in mains with stagnant water; living organisms that protect bacteria but that may release bacteria into the drinking water when mains are disturbed; and sewage cross-connections. In addition to bacterial contamination, corrosion by-products such as lead and copper can also deteriorate water quality and it is well established that disinfection by-products (DBPs) formed during treatment increase with time in the distribution system.

Distribution systems are generally designed to ensure hydraulic reliability, which includes adequate water quantity and pressure for fire flow as well as domestic and industrial demand. To meet these demands, extensive storage is usually incorporated in system design, which results in long residence times. Although these storage facilities can play a major role in providing hydraulic reliability for fire fighting and customer demand, they may also serve as vessels for complex chemical and biological changes that can result in the deterioration of water quality. These changes include the loss of disinfectant residuals and possible increases in the levels of microorganisms in the system.

The American Water Works Association Research Foundation (AWWARF) estimated that there is 880,000 miles of underground distribution piping in the United States with a replacement value of $348 billion. Much of this has been installed since World War II and is believed to be in good condition. However, about 27 percent is unlined cast iron and steel pipe that is judged to be in only fair or even poor condition. Older utilities have a predominance of aging infrastructure in poor condition that will require accelerated repair or replacement. Unfortunately, at the current rate of replacement, most utilities will only be able to replace a given pipe every 200 years. In its January 1997 Drinking Water Infrastructure Needs Survey, USEPA estimated that US drinking water utilities will need to spend $138.4 billion on infrastructure over the next 20 years in order to meet the requirements of the SDWA. The survey projected that 56 percent of this amount would be needed for transmission and distribution system installation and replacement.

There are potential health consequences resulting from the deterioration of the structural integrity of distribution systems in the US, and from current design and operations policies. Since 1993, nearly 850 community water systems have been ordered to issue public boil water advisories because of bacterial contamination from all causes. This includes highly publicized distribution system problems in a number of large cities, such as New York and Washington D.C. Distribution system failures in two small Missouri cities, Cabool and Gideon, are believed to be linked to hundreds of cases of intestinal illness and 13 deaths.

If good water quality is to be preserved, greater attention must be given to the design, operation, and maintenance of distribution systems. This includes continual improvement in capabilities to model and predict water quality in distribution systems, understanding the role of biofilms in effecting microbial activity in water, improved tank designs that promote mixing and avoid zones of low disinfectant residual, and improved techniques to manage corrosion by-product and disinfectant by-product formation in distribution systems. It is hoped that the techniques, procedures, and information contained in this book will help achieve that goal.

Acknowledgments

The research and studies described in this book took place over a span of more than 10 years and could not have been completed without the advice, assistance, and encouragement of many colleagues and associates. The authors gratefully acknowledge some of the individuals who provided this support, including Richard Males, PhD; Benjamin Lykins; James Goodrich, PhD; and Rolf Deininger, PhD; all of whom have been involved in all phases of this work and have continually encouraged our efforts. Thanks to Lewis Rossman, PhD, for his outstanding efforts in developing and validating EPANET; Harry Borchers, Judy Coyle, and David Milan, who were instrumental in implementing the first field studies ever conducted to demonstrate the potential for water quality modeling; Alan Hess, Ken Skov, and Darrell Smith, who provided many useful ideas to the authors and were deeply involved in validating many of the concepts described in this book; Gayle Smalley, who assisted in the first field application of water quality modeling for total trihalomethane propagation; Paul Boulos, PhD, who provided exceptional insight into many of the theoretical aspects of water quality modeling and, along with John Vasconcelos, PhD, field tested and applied EPANET to a broad cross section of water utilities in the United States; and to John Hill and Fred Angulo, PhD, for their assistance in applying EPANET for the first time to a waterborne outbreak scenario.

The authors extend appreciation to Jean Lillie, Sandra Taylor, Toni Frey, Steven Waltrip, and Nance Frazier for their assistance in preparing this manuscript.

Finally, we gratefully acknowledge our families, especially our wives, who have been so supportive and patient during the many hours spent at home and in the field developing, testing, and validating the concepts described in this book.

Distribution System
Water Quality

Drinking water treatment in the United States has played a major role in protecting public health by reducing waterborne disease. For example, the typhoid death rate for one year in the 1880s was 158 per 100,000 in Pittsburgh, Pa. By 1935, the typhoid death rate had declined to 5 per 100,000. Such reductions in waterborne disease outbreaks were brought about by the use of sand filtration, disinfection, and the application of drinking water standards (Clark, Ehreth, and Convery 1991).

This book discusses the use of water quality models and their potential for enhancing understanding of the factors that affect water quality in distributed water. The focus is primarily on research conducted by the authors, although the work of other investigators will be discussed where appropriate. The major elements involved in water quality modeling are outlined and the development and application of water quality models discussed. The results of applying these models, the development of storage tank models, and the effects of storage tanks on water quality in distribution systems will be discussed. Current research, including the development of EPANET, a state-of-the-art hydraulic and water quality model developed by USEPA, and its application to case studies will also be presented.

Federal Regulations

Concerns about waterborne disease and uncontrolled water pollution resulted in a dramatic increase in federal water quality legislation between 1890 and 1970. Even though significant advances were made to eliminate waterborne disease outbreaks during that period, the focus of drinking water concerns began to change. By the 1970s, more than 12,000 chemical compounds were known to be in commercial use and many more were being added each year. Many of these chemicals cause contamination of groundwater and surface water and are known to be carcinogenic and/or toxic. The passage of the Safe Drinking Water Act (SDWA) of 1974 was a reflection of concerns about chemical contamination.

The SDWA of 1974 and its amendments of 1986 required the US Environmental Protection Agency (USEPA) to establish maximum contaminant level goals (MCLGs) for each contaminant that may have an adverse health effect. Each goal is required to be set at a level at which no known or anticipated adverse effects on health occur, thereby allowing an adequate margin of safety (Clark, Adams, and Miltner 1987). Maximum contaminant levels (MCLs) or enforceable drinking water standards must be set as close to MCLGs as feasible.

The 1996 amendments to the SDWA nearly doubled the text of the act and provided for several new programs and initiatives, including the following: encouraging capacity development on the part of drinking water utilities, guidelines for water conservation, a requirement for utilities to increase efforts at informing their customers as to their status in complying with drinking water regulations, establishing an infrastructure grants program, emphasizing the need to bring small systems into compliance, and creating a whole new emphasis on source water protection.

The SDWA poses a massive challenge to the US drinking water industry. A large number of regulations will be implemented over a short period and the water utility industry will have difficulty in meeting the various rules and regulations established under the SDWA and its amendments.

Until recently, the emphasis in understanding, interpreting, and implementing the SDWA and its regulations has been on treated water. There is substantial evidence, however, that many factors can cause water quality to deteriorate between the treatment plant and the point of consumption. Some of these factors include the following: chemical and biological quality of source water; effectiveness and efficiency of treatment processes; adequacy of the treatment facility, storage facilities, and distribution system; age, type, design, and maintenance of the distribution network; and quality of treated water (Clark and Coyle 1990).

Most of the regulations established under the SDWA were promulgated with little understanding of the effect that the distribution system can have on water quality. However, the SDWA has been interpreted as meaning that some MCLs shall be met at the consumers' tap, and thus involves the entire distribution system.

Safe Drinking Water Act regulations that emphasize system monitoring include the Surface Water Treatment Rule (SWTR), the Total Coliform Rule (TCR), the Lead and Copper Rule, and the Trihalomethane Regulation. Both the SWTR and the TCR specify treatment and monitoring requirements that must be met by all public water suppliers.

The SWTR requires that a detectable disinfectant residual be maintained at representative locations in the distribution system to provide protection from microbial contamination. The TCR regulates coliform bacteria that are used as "surrogate organisms" to indicate whether or not system contamination is occurring. Monitoring for compliance with the Lead and Copper Rule is based entirely on samples taken at the consumer's tap. The current standard for trihalomethanes (THMs) is 0.1 mg/L for systems serving more than 10,000 people but the anticipated Disinfectant–Disinfection By-products (D–DBP) Rule may impose the current (or a reduced) THM level on all systems. This regulation also requires monitoring and compliance at selected monitoring points in the distribution system. Some of these regulations may, however, provide contradictory guidance. For example, the SWTR and TCR recommend the use of chlorine to minimize risk from microbiological contamination. However, chlorine or other disinfectants interact with natural organic matter (NOM) in treated water to form disinfection by-products (DBPs). Raising the pH of treated water will assist in controlling corrosion but will increase the formation of THMs.

Aging Infrastructure

Distribution systems are frequently designed to ensure hydraulic reliability, which includes adequate water quantity and pressure for fire flow and domestic and industrial demand. In order to meet these goals, large amounts of storage are usually incorporated into system design. This results in long residence times, which in turn, may contribute to water quality deterioration.

Many water distribution systems in the US are approaching 100 years old, and an estimated 26 percent of distribution system pipe is unlined cast iron and steel that is in poor condition. At current replacement rates for distribution system components, it is estimated that a utility will replace a pipe every 200 years (Kirmeyer, Richards, and Smith 1994).

Conservative design philosophies, an aging water supply infrastructure, and increasingly stringent drinking water standards have resulted in concerns over the viability of drinking water systems in the US. Questions have been raised about the structural integrity of drinking water systems in the US, as well as their ability to maintain water quality from the treatment plant to the consumer.

The Distribution System as a Barrier to Waterborne Disease

Developing an effective drinking water infrastructure is essential to public health, spurs productivity and profitability, increases the tax base, and creates skilled jobs in construction, engineering, and manufacturing. It has been estimated that 57,000 jobs are created for every $1 billion invested in distribution infrastructure (Kirmeyer, Richards, and Smith 1994). On the other hand, improperly designed and maintained systems can be an economic liability.

An infrequently considered factor that may influence water quality in a distribution system is the effect of mixing water from different sources. Water distribution systems frequently draw water from multiple sources, such as a combination of wells and/or surface sources. The mixing of waters from different sources that takes place within a distribution system is a function of complex system hydraulics (Clark, Grayman, and Males 1988; Clark, Grayman, and Goodrich 1991; Clark et al. 1991).

Much has been written about the multiple barrier concept and its application in providing protection against waterborne disease (Clark, Goodrich, and Wymer 1993). The *multiple barrier concept* refers to the use of conventional treatment in combination with disinfection to provide safe and aesthetically acceptable drinking water. Relatively little has been written, however, about the distribution system's role as a final barrier against waterborne disease.

DISINFECTION FAILURES

One indication of the need to consider the factors that affect water quality deterioration in the distribution system is the increasing number of reported incidents in which the occurrence of coliform bacteria has been documented in the presence of free chlorine residuals (Clark et al. 1991). In one reported case, routine bacteriologic monitoring of a distribution system serving 350,000 people in Connecticut showed elevated total coliform counts. The mean counts were 12.1 coliforms/100 mL as measured against a standard of 1 coliform/100 mL. The distribution system received water from four reservoir systems and five well fields. Water from three of the reservoir systems was filtered and chlorinated. Water from the other reservoir and from the well fields was chlorinated but not filtered. Free chlorine residuals were maintained at 0.6–1.0 mg/L in treatment effluents from surface sources, 0.2–0.4 mg/L at the well fields, and 0.2–0.5 mg/L at points in the distribution system. An aggressive water quality monitoring program coupled with a series of field research projects directed at solving these problems has prevented the situation from recurring (*Morbidity and Mortality Weekly Report* 1985).

An illustration of the water quality problems associated with failures in distribution is a recent study aimed at determining the movement of a waterborne contaminant found in the Cabool, Mo., distribution system. During the period Dec. 15, 1989, to Jan. 20, 1990, residents and visitors to Cabool, population 2,090, experienced 240 cases of diarrhea and 6 deaths. The organism *Escherichia coli* serotype 0157:H7, associated with the feces of healthy dairy cattle, was isolated in many of the stool samples of ill people.

An investigation performed by the Centers for Disease Control and Prevention (CDC) with assistance by USEPA's Water Supply and Water Resources Division (formerly the Drinking Water Research Division), concluded that the illness was caused by waterborne contaminants that entered the distribution system through a series of line breaks and meter replacements that occurred during unusually cold weather. This conclusion was based on statistical studies performed by CDC and corroborated by water quality modeling performed by USEPA. The study provides an example of how a water quality model can be

4

used to study contaminant propagation in a distribution system (Clark et al. 1991; Geldreich et al. 1992).

Recent experience has demonstrated that there may be problems with drinking water systems in the United States. Table 1-1 summarizes some of the more notable recent water supply problems (Clark et al. 1994). Many more communities than those listed in Table 1-1 have been placed on boil water orders or experienced coliform MCL violations over the last 3–4 years. In Milwaukee, Wis., it is estimated that a cryptosporidiosis outbreak infected more than 400,000 water consumers. The number of deaths associated with this outbreak is estimated at near 100 immune-compromised individuals. Boil water orders in the Manhattan borough of New York City and Washington, D.C., have drawn attention to drinking water problems in major metropolitan areas.

Another example of infrastructure failure occurred in Gideon, Mo., in November 1993 when, out of a population of 1,000, 400–500 people contracted *Salmonella typhimurium*. The *Salmonella* outbreak contributed to the death of seven of these individuals. It is presumed that bird droppings contaminated storage tanks. As with Cabool, the city used a nondisinfected groundwater. In both situations, deaths were among the elderly in nursing homes. One consequence of these investigations is the finding that immuno-compromised people (elderly and young) are often less capable of surviving waterborne illness than younger, healthier people (Clark et al. 1994).

A USEPA team was also invited by the CDC to assist in evaluating the waterborne disease potential associated with a recent cholera outbreak in Peru. The investigative team concluded that one of the major factors in this outbreak was the marginal condition of the distribution systems in the areas

Table 1-1   Recent water quality problems in the United States

City	Population Affected	Date of Onset	Cause of Problem	Results
Milwaukee, Wis.	800,000	Apr. 7, 1993	Cryptosporidiosis	Estimated 400,000 people ill
Washington, D.C.	1,000,000	Dec. 8, 1993	Turbidity violation	Boil water order
Boca Raton, Fla.	106,000	Jan. 30, 1991	Coliform violation	Boil water order
Talent, Ore.	3,000	May 22, 1992	Cryptosporidiosis	3,000 people ill
Carrollton, Ga.	18,000	Jan. 30, 1987	Cryptosporidiosis	13,000 people ill
Cabool, Mo.	2,090	Dec. 15, 1993	*E. coli* 0157:H7	6 people died, 85 were sick with bloody diarrhea
Gideon, Mo.	1,009	Nov. 29, 1993	*Salmonella typhimurium*	500 people ill, 7 people died
New York, N.Y. (Manhattan)	35,000	July 8, 1993	Coliform violation	Boil water order
Utica, N.Y.	135,000	November 1992	Coliform violation	State conference action plan

visited. It was observed that water leaving a treatment plant had adequate disin-fectant residuals that quickly disappeared within the distribution system. Sys-tems experienced intermittent operation, fluctuating pressure, frequent pipe breaks, high water losses, and had unplanned cross-connections. Little repair and maintenance was practiced by the utilities visited. It was the opinion of the investigative team that many of the waterborne disease problems associ-ated with cholera were related to improperly operated and poorly maintained distribution systems (Clark et al. 1993).

We should not be surprised that an increasing number of problems are found as water quality in drinking water distribution systems is more carefully examined. Part of the increased frequency in reported problems results from increasingly sensitive chemical and microbiological monitoring methods and aging infrastructure.

Another reason for increasing problems lies in the underlying design of distribution systems in most municipalities in the United States. Drinking water distribution systems are generally required to meet the standby or ready-to-serve requirements for fire fighting as well as domestic, industrial, and other normal water use (Fair and Geyer 1956). The National Fire Protection Association[*] governs the fire-fighting capacity of distribution systems. In order to satisfy the need for adequate capacity and pressure, most distribution systems use standpipes, elevated tanks, and storage reservoirs. Distribution systems are typically "zoned" due to a desire to maintain relatively constant pressures in the system or because of the way in which the system has expanded.

The effect of designing a system to maintain adequate fire flow and reli-ability can result in long transit times between the treatment plant and the consumer. However, long travel time and low velocities may be detrimental to meeting drinking water standards. Long residence times lead to the maximum formation of DBPs, loss of disinfectant residuals, and formation of biofilm. Clearly, the objective of meeting public safety and reliability requirements may be contrary to sound public health practice (Clark et al. 1993).

Even medium-sized water utilities may have thousands of miles of pipes constructed from various types of materials, ranging from new lined pipes to unlined pipes that are more than 50 years old. Over time, biofilms and tuber-cles attached to pipe walls can result in a significant loss of disinfectant resid-ual and, thereby, adversely affect water quality.

It has also been found that chlorine can be lost through both the interac-tion with NOM in raw water and with the pipe walls themselves in transport-ing finished water. This mechanism for loss of chlorine may be even more serious than long residence times in tanks. Pipe wall demand (the demand for chlorine at the pipe wall), possibly due to biofilm and tubercles, may be using up the chlorine very rapidly in a distribution system, and pipe cleaning or replace-ment may be required to minimize these problems. Maintaining adequate levels

---

[*] One Batterymarch Park, Quincy MA 02269-9101.

of disinfectant residual may require both clean pipes and intensive treatment (Clark et al. 1993).

If there is a break in the distribution system or a contaminant enters the system through some kind of unanticipated event, contaminants can move back and forth in the pipes under varying demand and hydraulic scenarios. This effect was illustrated in research conducted on contaminant propagation by the authors in conjunction with the North Penn Water Authority located north of Philadelphia, Pa., and confirmed during the waterborne disease outbreak in Cabool, Mo. (Geldreich et al. 1992).

The distribution portion of a water supply system also generally represents the largest expenditure (more than 80 percent of capital costs) for nearly all water utilities. Typically the replacement value for the distribution system in a medium-sized utility may be billions of dollars (Clark and Stevie 1978).

It is difficult to study the problems of system design and effects of long residence times in a full-scale system. Constructing specially designed pipe loops is one approach to simulating full-scale systems, however, properly configured and calibrated mathematical models can be used to study water quality issues as well. Such models may also be used to assess various operational and design decisions, determine the impacts resulting from the inadvertent introduction of a contaminant into the distribution system, and assist in the design of an improved water quality sampling program.

Water Quality Deterioration in Distribution Systems

There are many opportunities for water quality to change as it moves between the treatment plant and the consumer. Figures 1-1 and 1-2 illustrate some of the transformations that take place in the water flow through the pipe (bulk water phase) and at the pipe wall. Cross-connections, failures at the treatment barrier, and transformations in the bulk phase can all degrade water quality. Corrosion, leaching of pipe material, and biofilm formation and scour can occur at the pipe wall to degrade water quality.

Unlike hydraulic modeling of distribution systems, water quality modeling is a relatively recent development dating back only to the early 1980s. Water quality models can be used to trace the movement of contaminants or any element in distribution system water. These models may also be used to calculate the age of the water within the distribution system or to determine the source(s) of water throughout the system. Water quality models generally use the results of hydraulic models and, like hydraulic models, may be based on either steady state or extended period simulation (EPS). Contaminants may be treated as either conservative (does not degrade) or as experiencing decay or growth as they move through the system.

Many researchers have investigated how water deteriorates once it enters the distribution system. It has been documented that bacteriological quality

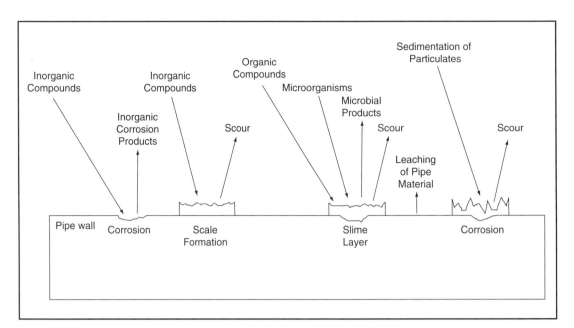

Figure 1-1   Schematic of chemical and microbiological transformations at the pipe wall

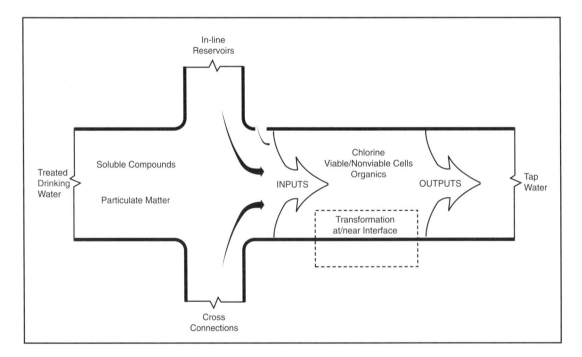

Figure 1-2   Microbiological and chemical transformations in drinking water

changes can cause taste-and-odor problems, discoloration, slime growths, and economic problems, including corrosion of pipes and biodeterioration of materials (Water Research Centre 1976). Bacterial numbers tend to increase during distribution and are influenced by a number of factors, including bacterial quality of the finished water entering the system, temperature, residence time, presence or absence of a disinfectant residual, construction materials, and availability of nutrients for growth (Geldreich et al. 1972; LeChevallier, Babcock, and Lee 1987; Maul, El-Shaarawi, and Block 1985a and b).

The relationship of bacteriological quality to turbidity and particle counts in distribution water was studied by McCoy and Olson (1986). One upstream and one downstream sampling site in each of three distribution systems (two surface water supplies and one groundwater supply) were sampled twice per month over a one-year period. Degradation of bacterial water quality was shown to be the result of unpredictable intermittent events that occurred within the system.

Figures 1-3, 1-4, and 1-5 illustrate some of the complex chemical and microbiological changes that take place in distribution networks (Clark et al.

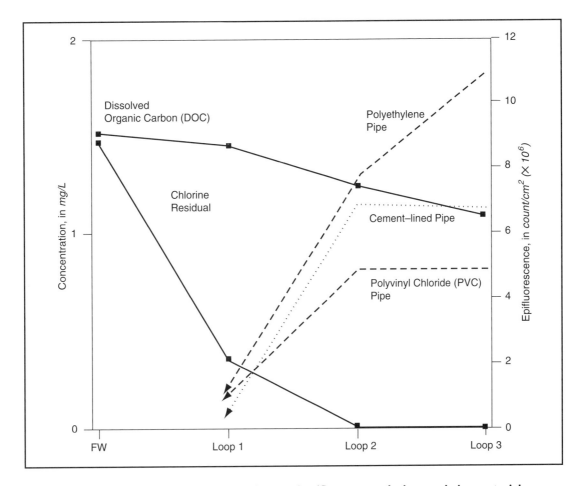

Figure 1-3 Chlorine, dissolved organic carbon, and epifluorescence by loop and pipe material

1994). These data are from a unique pilot facility in Nancy, France. The pilot facility consisted of two pilot plants operating in parallel, each feeding a separate simulated distribution system. The source of water for the pilot plant was a nondisinfected raw surface water. Each simulated system consisted of three pipe loops in series. Each of the loops had a one-day residence time.

It is difficult at this time, given our current level of knowledge, to predict changes in water quality during transport in a distribution network. However, as can be seen from these data, as the water moves through the loops there is a tendency for the disinfectant residual to drop and various measures of biological activity in the bulk phase to increase.

Based on the study in Nancy, Figure 1-3 illustrates the decrease in dissolved organic carbon (DOC) that took place as water traveled through the

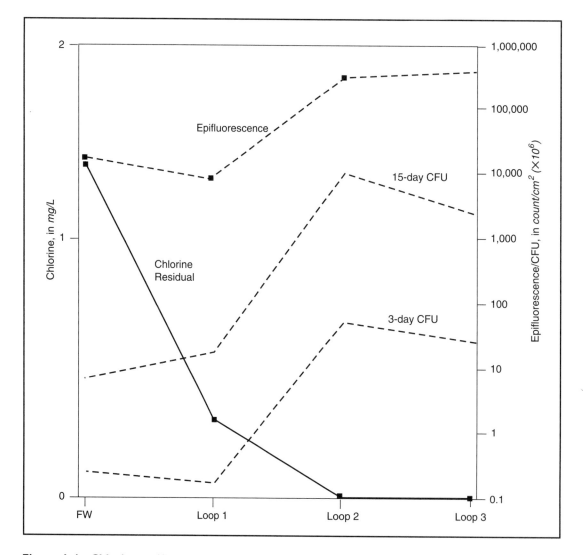

Figure 1-4   Chlorine, epifluorescence, and colony-forming units (CFU) by loop

pilot network. This supports the assumption that DOC was used for biological growth. The biofilm buildup increased with residence time in the loop, and increased dramatically after disinfectant residuals disappeared. This is consistent with results shown by LeChevallier, Babcock, and Lee (1987). Polyethylene coupons sustained the highest biofilm buildup, perhaps due to materials in the pipe that encourage biofilm growth. There are clearly differences in biofilm formation based on the type of pipe material.

Figure 1-4 shows the steady loss of disinfectant residual through the pipe loops. The chlorine residual dropped most rapidly because of its reaction with organic material in the bulk phase of the water and with the pipe wall material. As the residuals dropped, the organisms in the bulk water increased dramatically.

Figure 1-5 shows THM changes that took place across the pipe loops for postdisinfection with chlorine. The drop in trihalomethane formation potential (THMFP) was no doubt related to the consumption of DOC.

As shown in Figure 1-5, the disinfectant residual dropped rapidly and was virtually gone by the end of the second loop. Instantaneous THMs exceeded THMFP at the end of loop 2 and instantaneous THMs continued to increase

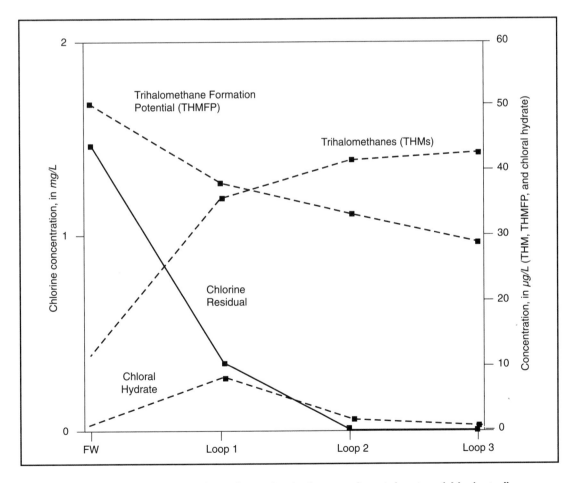

Figure 1-5   Change in trihalomethane formation in the experimental system (chlorinated)

through the third loop. This might be explained by the increase and then decrease in chloral hydrate concentrations between the treatment plant and the end of the third loop. Chloral hydrate may be converted to THMs in the absence of chlorine. These results are consistent with other research that has shown this same phenomenon (Stevens, Moore, and Miltner 1989).

References

Clark, R.M., and R.G. Stevie. 1978. Meeting the Drinking Water Standards, Safe Drinking Water: Current and Future Problems. In *Proceedings of a National Conference, Resources for the Future*. Washington, D.C.: Resources for the Future.

Clark, R.M., J.Q. Adams, and R.M. Miltner. 1987. Cost and Performance Modeling for Regulatory Decision Making. *Water*, 28(3):20–27.

Clark, R.M., W.M. Grayman, and R.M. Males. 1988. Contaminant Propagation in Distribution Systems. *Jour. Environ. Eng.*, 114(2).

Clark, R.M., and J.A. Coyle. 1990. Measuring and Modeling Variations in Distribution System Water Quality. *Jour. AWWA*, 82(8):46.

Clark, R.M., W.M. Grayman, and J.A. Goodrich. 1991. Water Quality Modeling: Its Regulatory Implications. In *Proc. AWWARF/EPA Conf. on Water Quality Modeling in Dist. Systems*. Cincinnati, Ohio: USEPA.

Clark, R.M., D.J. Ehreth, and J.J. Convery. 1991. Water Legislation In the U.S.: An Overview of The Safe Drinking Water Act. *Toxicology and Industrial Health*, 7(516):43–52.

Clark, R.M., W.M. Grayman, J.A. Goodrich, R.A. Deininger, and A.F. Hess. 1991. Field Testing Distribution Water Quality Models. *Jour. AWWA*, 83(7):67.

Clark, R.M., J.A. Goodrich, and L.J. Wymer. 1993. Effect of the Distribution System on Drinking Water Quality. *Jour. Water Supply Research and Technology–AQUA*, 42(1):30–38.

Clark, R.M., W.M. Grayman, R.M. Males, and A.F. Hess. 1993. Modeling Contaminant Propagation in Drinking Water Distribution Systems. *Jour. Environ. Eng.*, 119(2):349–364.

Clark, R.M., J.E. Hill, J.A. Goodrich, J.A. Barnick, and F. Abdesaken. 1994. Impact of Tanks and Reservoirs On Water Quality In Drinking Water Distribution Systems: Regulatory Concerns. In *Proceedings 1994 AWWA Annual Conference*. Denver, Colo.: American Water Works Association.

Clark, R.M., B.W. Lykins, Jr., J.C. Block, L.J. Wymer, and D.J. Reasoner. 1994. Water Quality Changes in a Simulated Distribution System. *Jour. Water Supply Research and Technology–Aqua*, 43(6):263–277.

Fair, G.M., and J.C. Geyer. 1956. *Water Supply and Waste Water Disposal*. New York: John Wiley and Sons, Inc.

Geldreich, E.E., H.D. Nash, D.J. Reasoner, and R.H. Taylor. 1972. The Necessity of Controlling Bacterial Populations in Potable Water: Community Water Supply. *Jour. AWWA*, 64(9):596–602.

Geldreich, E.E., K.R. Fox, J.A. Goodrich, E.W. Rice, R.M. Clark, and D.L. Swerdlow. 1992. Searching for a Water Supply Connection in the Cabool, Missouri, Disease Outbreak of *Escherichia Coli* 0157:H7. *Water Research*, 26(8):1,127–1,137.

Kirmeyer, G.J., W. Richards, and C.D. Smith. 1994. *An Assessment of the Condition of North American Water Distribution Systems and Associated Research Needs*. Denver, Colo.: American Water Works Association Research Foundation and American Water Works Association.

LeChevallier, M.W., T.M. Babcock, and R.G. Lee. 1987. Examination and Characterization of Distribution System Biofilms. *Applied and Environmental Microbiology*, 53:2,714–2,724.

Maul, A., A.H. El-Shaarawi, and J.C. Block. 1985a. Heterotrophic Bacteria in Water Distribution Systems. I. Spatial and Temporal Variation. *The Science of the Total Environment*, 44:201–214.

———. 1985b. Heterotrophic Bacteria in Water Distribution Systems. II. Sampling Design for Monitoring. *The Science of the Total Environment*, 44:215–224.

McCoy, W.F., and B.H. Olson. 1986. Relationship Among Turbidity Particle Counts and Bacteriological Quality Within Water Distribution Lines. *Water Research*, 20:1,023–1,029.

Morbidity and Mortality Weekly Rept., 34:10:142 (Mar. 15, 1985). Detection of Elevated Levels of Coliform Bacteria in a Public Water Supply, Connecticut.

Stevens, A.A., L.A. Moore, and R.J. Miltner. 1989. Formation and Control of Non-Trihalomethane Disinfection By-Products. *Jour. AWWA* 84:54–60.

Water Research Centre. 1976. Deterioration of Bacteriological Quality of Water During Distribution. Notes on Water Research No. 6. Water Week, Vol. 3, No. 10, 1 (1994).

Modeling Distribution Systems

The use of mathematical models for the analysis of water distribution systems was first proposed in the 1930s by Hardy Cross (1936). Since that time, as illustrated in Figure 2-1, models have evolved from the early manual Hardy Cross network flow analyses, to the development and wide availability of computer-based hydraulic network models in the 1970s and 1980s, to the emergence of distribution system water quality models later in the 1980s. Modeling and mapping systems are now being integrated into comprehensive, user-friendly systems for analyzing and displaying hydraulic and water quality parameters in a distribution system. As will be discussed later in this book, water quality models have reached operational status, but research and development continues to further the understanding of the processes taking place in the distribution system and to translate this understanding into usable tools.

In applying models to a water distribution system, the degree of temporal (over time) variation and the specific issues that are being studied determine the types of models that are most applicable. For example, *steady state modeling* represents external forces as constant in time and determines solutions that would occur if the system were allowed to reach equilibrium (Wood 1980a). In *dynamic modeling*, demands and supplies are allowed to vary and the resulting solution that varies over time is determined (Clark, Grayman, and Males 1988; Clark et al. 1988b).

In both steady state and dynamic modeling, a distribution system is represented by a link–node network (i.e., pipes are represented as links, and junctions of pipes, wells, tanks, or starts of pipes as nodes). Hydraulic models are used to determine flows and velocities in links.

Water quality models are used to determine variations in the concentrations of a contaminant throughout the distribution system. The hydraulic and water quality models may be an integrated package or the results from a hydraulic model may be input to a water quality model for subsequent analysis.

Types of Models

A variety of computer-based mathematical models of water distribution systems have been developed and used by the water utility industry to assess the movement and fate of contaminants within the distribution system. Such models may be divided into three general categories:

- *hydraulic models*, which simulate the flow quantity, flow direction, and pressure in the system

- *steady state water quality models and flow-tracing models*, which determine the movement of contaminants, including their flow paths and travel times, through the network under steady state operational and demand conditions

- *dynamic water quality models*, which simulate the movement and transformation of substances in the water under conditions that vary over time

Two related technologies, which are not discussed in this book, are

- *optimization models incorporating water quality*, which examine a wide range of operational and/or design variables and select the best solution based on a stated objective function and specified constraints

- *analysis/display methods*, which perform various analyses and/or display the results of water quality sampling or modeling

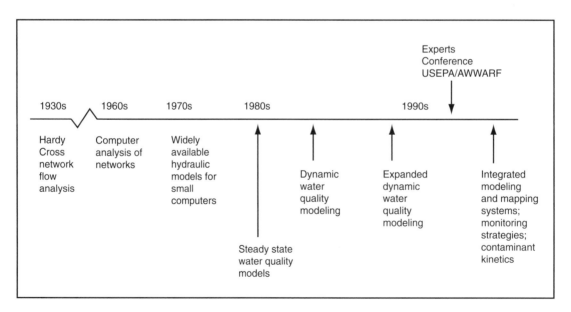

Figure 2-1   Historical development of water distribution system modeling

Each type of model serves a particular purpose in assessing distribution system water quality and is essential in investigating water quality issues in a distribution system.

HYDRAULIC MODELS

As mentioned previously, mathematical methods for analyzing the flow in networks have been in use for over half a century (Cross 1936). Computer-based models for performing this type of analysis were first developed in the 1950s and 1960s and were greatly expanded and made more available in the 1970s and 1980s. Currently, dozens of such models are readily available for use on systems ranging from personal computers (Wood 1980b) to supercomputers (Sarikelle, Chuang, and Loesch 1989).

Hydraulic models developed to simulate flow and pressures in a distribution system, either under steady state conditions or under time varying demand and operational conditions, are generally referred to as *extended period simulation* (EPS) models. Hydraulic models may also incorporate optimization components to aid the user in selecting system parameters that result in the best match between observed system performance and model results (Gessler and Walski 1985). (For theory and application of such hydraulic models, refer to Gessler and Walski [1985] and American Water Works Association [1989]).

Hydraulic models also provide flow information used in distribution system water quality models. Hydraulic models and water quality models can be tightly bundled into a single entity, or a stand-alone hydraulic model can be used to generate a file containing hydraulic flow conditions that can be used in a stand-alone water quality model. Both techniques are in common use today.

STEADY STATE WATER QUALITY MODELS

The concept of using models to determine the spatial pattern of water quality in a distribution system using sources of differing quality was suggested in a study of slurry flow in a pipe network (Wood 1980a). This study presented an extension to a steady state hydraulic model in which a series of simultaneous equations are solved for each node.

A similar formulation was later used in a 166-link representation of the Alameda County, Calif., Water District with three sources of water of differing hardness (Chun and Selznick 1985; Metzger 1985).

In a generalization of this formulation, Males et al. (1985) used simultaneous equations to calculate the spatial distribution of concentration, travel times, costs, and any other variable that could be associated with links or nodes. This model, called SOLVER, was a component of the Water Supply Simulation Model (WSSM), an integrated database management, modeling, and display system (Clark and Males 1986).

An alternative steady state "incremental" solution was introduced by Clark et al. (1988b) for calculating spatial patterns of concentrations, travel times, and the percentage of flow from sources. In this approach, links are hydraulically ordered starting with source nodes and progressing through the network until all nodes and links are addressed.

Alternative methodologies for predicting water quality and determining the source of delivered flow under steady state conditions was investigated by Wood and Ormsbee (1989). An iterative cyclic procedure was found to be both an effective and efficient method. This procedure was similar to the incremental solutions described previously for networks that they term as *source dependent* (i.e., networks in which the nodes can be hydraulically sequenced starting from sources). However, for nonsource-dependent networks, which are rare, their algorithm iterates until a unique solution is found.

DYNAMIC WATER QUALITY MODELS

Though steady state water quality models proved to be useful tools for investigating the movement of a contaminant under constant conditions, the need for models that represent the dynamics of contaminant movement was recognized. In the mid-1980s, several models that simulate the movement and transformation of contaminants in a distribution system under time varying conditions were developed and applied. Three such models were initially introduced at the American Water Works Association (AWWA) Distribution Systems Symposium in 1986 (Clark et al. 1986; Liou and Kroon 1986; Hart, Meader, and Chiang 1986).

Grayman, Clark, and Males (1988) developed and applied a water quality simulation model that uses flow results previously generated by a hydraulic model and a numerical scheme to route conservative (i.e., concentration does not degrade with time) and nonconservative (i.e., concentration changes with time) contaminants through a network. In this model, each pipe link is represented as a series of "sublinks" and "subnodes" with the length of each sublink selected to approximate the distance that a contaminant will travel during each time step. The number of sublinks vary with the velocity of flow in a link (Grayman, Clark, and Males 1988). Kroon and Hunt (1988) developed a similar numerical model originally implemented on a mini computer and now available for use on a PC-based workstation (Liou and Kroon 1986). This model is directly tied to a hydraulic model and generates both tabular and graphical output displaying the spread of contaminants through a network. Hart, Meader, and Chiang (1986) developed a model using the GASP IV simulation language.

Model Verification

The types of models described have been applied in several research studies. In a cooperative agreement between the US Environmental Protection Agency (USEPA) and the North Penn Water Authority, a series of water quality models were developed and applied in order to study the movement of water through the distribution system (Clark and Coyle 1990). In this system, trihalomethanes (THMs) were higher in the surface water source, while hardness was higher in the groundwater sources. Modeling was used to trace the movement of these constituents and, ultimately, to alter the operation in order to minimize water quality variation. The temporal and spatial variability of these constituents was quite significant and, as has been confirmed in many other studies, the dynamics of the contaminant movement in the system exceeded the general expectations of the water utility (Clark, Grayman, and Males 1988).

In a cooperative agreement involving the AWWA Research Foundation (AWWARF), the South Central Connecticut Regional Water Authority (SCCRWA), Montana State University, and Stoner Associates, water quality models were developed and applied to the Saltonstall service area of the SCCRWA to study the chlorine residual in the system and its impact on bacterial regrowth (Characklis 1988). These models were later applied to the Philadelphia Suburban Water Company's Great Valley Division to determine the percentage of water from various groundwater and surface sources reaching various parts of the system (Kroon 1990).

In another cooperative agreement, USEPA and the University of Michigan applied water quality models to the Cheshire service area of the SCCRWA. A highly skeletonized representation (i.e., network contains only major transmission lines) of the full SCCRWA system was developed and the effects of skeletonization on water quality were studied (Clark et al. 1990). Both fluoride (used as a tracer) and chlorine residual were modeled in this study. This study reinforced many of the previous studies in illustrating the significant dynamics in water quality within a distribution system resulting in large variations over the network and over time in concentrations (Clark et al. 1990).

In the 1980s distribution system water quality models began to emerge. In the early 1990s modeling and mapping systems were integrated into comprehensive, user-friendly systems for analyzing and displaying hydraulic and water quality data. Such models may be used to assess alternative operational and design decisions under normal or emergency conditions. They are also frequently used to study likely design needs in the water system in response to future growth.

CALIBRATION

Calibration is an important part of the "art" of modeling drinking water distribution systems. *Model calibration* is the process of adjusting model input data (or, in some cases, model structure) so that the simulated hydraulic and water quality outputs sufficiently mirror observed field data. One way of viewing calibration is to think of a television screen showing observed and predicted values with knobs available to adjust the predicted values. Calibration is the process of adjusting the knobs so that the predicted value provides the closest estimate to the actual values. Calibration can be difficult, costly, and time-consuming.

The extent and difficulty of calibration is minimized by developing an accurate representation of the network and its components. Accurate system performance and operations data should be collected to validate the network simulations. A traditional technique for calibration is to use fire flow pressure measurements. Pressures and flow in isolated pipe sections are measured in the field and the pipe friction factors are adjusted to reflect the data.

Another method is to use water quality tracers. Naturally occurring or added chemical tracers may be measured in the field and the results used to calibrate hydraulic and water quality models.

Calibration should be an ongoing process in which the degree of accuracy of the simulations are continually being improved (Cesario 1995).

Modeling Applications

In 1991 USEPA and AWWARF cosponsored the Water Quality Modeling in Distribution Systems conference to identify the state-of-the-art in water quality contaminant propagation modeling. The symposium identified the following 11 categories for water quality modeling research and application:

- model development
- chemical, physical, and biological processes
- modeling application
- water quality studies
- calibration of models
- regulatory implications
- monitoring strategies
- legal proceedings
- management and operational cases
- operation and design of storage tanks
- health effects

Although this agenda was established in 1991, it has become the template for research in the field of distribution system water quality modeling as followed by AWWARF and USEPA.

MODEL DEVELOPMENT

A number of investigators have developed or are developing water quality models (including dynamic and optimization models). Dynamic water quality models have been developed by several investigators to predict the propagation of contaminants in distribution systems. Optimization models provide a mathematical means of generating a large number of solutions and selecting the solution that best fills an objective within specified constraints. Within the water quality field, optimization has been used to select the best way to mix and treat waters of differing quality from different sources, and to schedule pumping from multiple sources to meet desired water quality goals at a minimum cost. Nonlinear optimization techniques have been used to determine the least-cost treatment and mixing solution in a distribution system with steady state conditions under various water quality and head constraints. Interactive computer programs have been developed to calculate optimal schedules for the pump stations and treatment plants in a water distribution system. The objective for the system is to deliver the required water at a minimum operating cost while maintaining the system operation within feasible ranges for water quality, reservoir levels, and system flows.

Because of the large amount of information generated by water quality modeling, methods for graphically displaying results from models are essential. Most operational models contain graphical modules that display results in the form of network plots or contour plots.

CHEMICAL, PHYSICAL, AND BIOLOGICAL PROCESSES

Operational distribution system models generally represent the physical, biological, and chemical processes that occur in a distribution system by simplified mathematical relationships. Typically, contaminants are treated as conservative substances or represented by a first-order decay function of the form

$$C_{t+\Delta t} = C_t \times e^{-(\Delta t \times k)}$$ (Eq 2-1)

Where:

C_t = concentration at time t, in mg/L

$C_{t+\Delta t}$ = concentration at time Δt, in mg/L

k = decay coefficient, in time^{-1}

This type of decay relationship has been applied in modeling chlorine and organics. Complete mixing is generally assumed at all nodes along with homogeneous conditions through the pipe cross section. Recent research has shown that there may be a considerable loss of chlorine associated with pipe wall reactions. In older ferrous pipe systems, the predominant mechanism of chlorine residual loss appears to be the corrosion reaction that can be described as a zero-order kinetic decay reaction. In coated or plastic pipe, biofilm uptake appears to be the predominant mechanism for chlorine loss at the well. It can be described by a first-order kinetic decay reaction (Vasconcelos et al. 1996).

Though not routinely included in the available operational models, there is considerable research needed in the area of physical, chemical, and biological processes in the distribution system.

APPLICATION

In order to apply a water quality model to a distribution system, it is necessary to develop a hydraulic representation of the system, represent the water quality sources and processes, calibrate the hydraulic and water quality model, and, finally, apply the model to study the operational or design parameters of concern. Several experimental water quality models have been developed, refined, calibrated, and assessed for viability.

WATER QUALITY STUDIES

Many of the early applications of water quality models have been in research studies in which the models were developed and refined. A water quality modeling package called PICCOLO has been applied in parts of Paris, France, to study chlorine concentrations. Construction of a water quality model of the city of Madrid, Spain, is under way and applications of models in Israel, the United Kingdom, and France are also being undertaken.

A unique series of small-scale applications of water quality models has been reported in which models have been used to represent detailed piping systems in apartments and industries to study such areas as lead buildup and bacterial growth in an ultrapure water system.

CALIBRATION

Calibration is the process by which model parameters are adjusted such that the model will produce predicted results similar to field results. The following four types of calibration procedures are commonly used with water quality models: (1) routine chemical and hydraulic measurements, (2) intensive surveys using unique "signature" concentrations corresponding to different sources, (3) intensive surveys using an injected tracer, and (4) small-scale, detailed field studies.

A popular method for calibrating water quality models is the use of fluoride as a tracer. Typically, for systems that routinely fluoridate their water and whose background level of fluoride is negligible, fluoride injection is discontinued for a period during which distribution system sampling occurs. Further sampling can then take place during the transient period when fluoride injection begins again. The largest-scale calibration study of this type was performed in the Cheshire Service Area of the SCCRWA distribution system and will be discussed in detail later. In this study, fluoride was turned off for a seven-day period and then restarted for another seven days. Intensive sampling occurred during this 14-day period. Before the study, water quality models were applied to the area to predict the fluoride movement to establish sites and timing for sampling. Over 2,000 grab samples were taken during the study. These samples were augmented by samples taken using an automated sampler receiving information from a fluoride probe. Following the sampling study, a water quality model was applied under the hydraulic conditions existing during the sampling period. The model was found to give good agreement with the sample results. Fluoride concentrations measured at the inlet and outlet of a storage tank in the Cheshire system were used to calibrate a model of the tank and to demonstrate the existence of short-circuiting of flow in the tank.

The effects of skeletonization of networks on the accuracy of hydraulic and water quality modeling has also been studied. These studies have concluded that skeletonization based on sound engineering judgment is a cost-effective means of representing complex networks but that water quality modeling results are generally more sensitive to skeletonization than are hydraulic model results.

REGULATORY IMPLICATIONS

There appears to be a synergistic relationship between regulation of water quality in the distribution system and water quality modeling. Much of the impetus and interest in water quality modeling has been spurred by the passage of the Safe Drinking Water Act and its amendments. The SDWA specifically requires that detectable levels of chlorine residual (0.1 mg/L or greater) be maintained and that heterotrophic plate counts (HPCs) not be over 500 cfu/100 mL. In particular, new regulations require that water quality meet specified standards at the tap.

Water quality complaints fall into several categories, including taste-and-odor and red water complaints. Tracing flows from the source, for example, can be helpful in responding to taste-and-odor complaints. Customers do not usually complain of lack of chlorine residuals or increasing HPCs but modeling can be used to solve these problems also.

Water quality modeling provides a tool that water utilities can use to estimate the effects at the tap of various treatment and delivery scenarios. Additionally, regulations regarding disinfection may cause significant changes in the disinfection processes. Water quality modeling provides a mechanism for determining the affect of changes in disinfection practices on delivered water.

MONITORING STRATEGIES

In the future, water quality models may be used to select monitoring station locations to maximize the informational content of sampling results. For example, models can be used to identify low or stagnant flow areas that might in turn be candidate sites for distribution network monitoring stations.

LEGAL PROCEEDINGS

Water quality modeling has been used extensively in legal proceedings involving the movement of a contaminant in a distribution system. However, because of the closed nature of the legal process, information on only a few studies has been published, thus, the lessons that could be learned from such applications cannot be shared. Typically, water quality models are used in legal proceedings as a means of showing the movement of water from a contaminated source to some point and to prove or disprove that the plaintiff has been damaged by the defendant. In theory, utilities would like to have water quality models but the time and effort involved make them a low priority until legal problems arise. In such cases, the utility has to backtrack and recreate the incident.

MANAGEMENT AND OPERATIONAL CASES

There are many potential uses of water quality models in the operation and management of a utility. As previously mentioned, models can be used to assess water quality (taste-and-odor) complaints made to a utility and to evaluate the impacts of the operation of a multisource system.

Water quality models can also be used in a variety of ways related to hydrant flushing programs, e.g., to indicate those mains that, because of low velocity, are most susceptible to deposition and most in need of flushing; to determine the volume of water required to adequately and economically flush a series of pipes; to isolate mains so that the discolored water generated during the flushing process does not spread to adjoining areas; and to predict the effects of a flushing program in saving time, energy, and water and avoiding customer complaints.

OPERATION AND DESIGN OF STORAGE TANKS

Water quality modeling techniques can be used to determine the affect of tank design and system operation on water quality in the distribution system. For example, a two-compartment model of a tank was developed to represent the mixing process in a tank in the Cheshire service area of the SCCRWA. This model was found to give a good approximation of the concentration of fluorides discharging from the tank during a fluoride sampling program. Using modeling, it was also determined that tanks can have a significant effect on the water quality in the distribution system. Water quality is very sensitive to tank location, mixing characteristics, and drawdown.

HEALTH EFFECTS

Contaminated water can have potentially catastrophic impacts on the health of customers. Water quality models can be applied to investigate the link between drinking water and waterborne disease.

Research Needs

The following research needs in the major areas related to modeling of water quality in distribution systems were identified at the 1991 USEPA/AWWARF conference:

- model structure and technology

- influence of tanks on water quality

- guidance and strategies for model usage

- legal and management implications

- system sampling and monitoring strategies

- chemical, physical, and biological issues

In addition, the conferees strongly endorsed the need for institutional structures and communications methods to ensure ongoing research and technology transfer in the field.

MODEL STRUCTURE AND TECHNOLOGY

Research is needed to investigate alternative methods of solving the basic quality modeling problem, including true dynamic models, as opposed to quasi-steady-state models; particle tracking methodologies; and other advanced mathematical techniques. The possibility of using supercomputers and/or parallel processing should be examined. Opportunities for incorporating graphical visualization and animation techniques into modeling should be explored. Research should be conducted into the possibility of extending the scope of existing models to include consideration of the total system (sources, treatment, and distribution) in quality models. Some basic assumptions need to be examined in more detail, in particular the assumption of complete mixing at junctions.

Research into alternative forms of operating control logic for dynamic hydraulic models to better simulate complex operating schemes and supervisory control and data acquisition (SCADA) systems is needed. More consideration of real-time operating issues is required. Appropriate methods of combining geographic information system (GIS), SCADA, and hydraulic and quality models need to be explored.

INFLUENCE OF TANKS ON WATER QUALITY

Extensive study of the role of tanks in distribution system water quality is needed. Studies are needed in regard to hydraulic and water quality issues and in improving existing models. Specific characteristics for study include inlet/outlet configuration and size, location of baffling, water quality changes within tanks, and methods of characterizing the initial age of water in tanks. The conflicts between design and operation for system reliability, fire protection, and water quality, in terms of location and sizing of tanks and system operation, need to be addressed. New methods for reducing tank-caused quality problems, such as disinfection of tank outflow or revision of tank design, also need to be explored.

GUIDANCE AND STRATEGIES FOR MODEL USAGE

Methods of determining when steady state modeling is sufficient for a given problem or when dynamic modeling must be pursued are needed, as are assessment methodologies to determine the need for highly detailed network models versus skeletonized models.

Guidelines and criteria for model calibration should be developed. Automated calibration techniques, perhaps using expert systems, need further examination. The problem of calibration at low flows and velocities is important. Methods for measuring the quality and acceptability of model results are needed to determine if one model or modeling calibration is superior to another.

LEGAL AND MANAGEMENT IMPLICATIONS

Additional study of legal and management issues associated with water quality models is required. Applications-oriented research into effective use of such models for emergency preparedness, counterterrorist planning, integration of water quality into operational decisions, and design of flushing programs would be valuable.

From a management and regulatory point of view, water quality models can be used to explore issues of spatial variations in risk of exposure and understand compliance mechanisms and effects of proposed regulations. Conflicting regulatory issues (e.g., the apparent opposition of goals for water quality and reliability) can be explored through modeling. Methods and demonstrations of model application in these areas are needed.

The legal implications of variations in water quality, particularly in terms of historical contamination of sources and determination of associated population exposure, are significant. Research into the required data, calibration, and modeling technology for such historical/forensic modeling is needed.

SYSTEM SAMPLING AND MONITORING STRATEGIES

Research is needed to explore optimal strategies for monitoring distribution systems. Methods of assessing the required number, timing, and location of samples for a given distribution system need to be developed. Methods using population exposure as the basis for assessment, rather than simply total flow, are required.

CHEMICAL, PHYSICAL, AND BIOLOGICAL PROCESSES

An area of very promising research is the examination of transformations within the pipe itself. There is a strong need to properly characterize chlorine residual decay rates, spatially and temporally, and to distinguish between contaminant decay rates in water and at pipe walls. Methods for field testing chlorine decay rates in pipes and reservoirs, and decay rate models for different disinfectants, pipe materials, holding time in tank, and demand conditions are needed.

Lead needs to be characterized as being in dissolved or particulate form. Fate and transport models need to take into account multicomponent mixtures and heterogeneous reactions. Formation and decay studies need to be carried out for TTHMs, brominated compounds, and disinfection by-products (DBPs). Total trihalomethanes and brominated compounds are subsets of disinfection by-products. Factors affecting development and control of biofilm need to be studied.

HUMAN EXPOSURE TO CONTAMINANTS

Studies of sediment transport in pipes are needed to determine possible sedimentation locations in a pipe. Biofilm-related studies, including TTHM spikes as an indication of sloughing and the degree to which flushing scours biofilm, are also needed. Identifying the impact of established biofilm/biocides on modeling water quality changes in the distribution system is needed, as is characterization of biocide activity at the pipe water interface in a biofilm.

Physical and chemical factors that set up stability of sediment sites should be identified, as should sediment concentration sites that affect microorganism inactivation in the pipe network. Research using field data to map the stability and diffusion of sediment in the pipe network environment that affects disinfectant effectiveness and biofilm control would be valuable. Settling in pipes, in terms of particle sizes, history of sediment accumulation, and settling rates in water of different chemical properties, needs to be characterized.

In the area of corrosion by-product formation, refined models are needed that consider chemical factors specific to water distribution systems. The factors that affect the corrosion of pipe material and the release of accumulation scales need to be better understood. Various corrosion models need to be evaluated in all respects for better use in evaluating specific corrosion problems.

References

American Water Works Association. 1989. *Distribution Network Analysis for Water Utilities.* Denver, Colo.: American Water Works Association.

Cesario, L. 1995. *Modeling, Analysis, and Design of Water Distribution Systems.* Denver, Colo.: American Water Works Association.

Characklis, W.G. 1988. *Bacterial Regrowth in Distribution Systems.* Denver, Colo.: American Water Works Association Research Foundation and American Water Works Association.

Chun, D.G., and H.L. Selznick. 1985. Computer Modeling of Distribution System Water Quality. *ASCE Spec. Conf. on Comp. Applic. in Water Res.*, New York: American Society of Civil Engineers.

Clark, R.M., and R.M. Males. 1986. Developing and Applying the Water Supply Simulation Model. *Jour. AWWA*, 78(8):61.

Clark, R.M., W.M. Grayman, R.M. Males, and J.A. Coyle. 1986. Predicting Water Quality in Distribution Systems. In *Proc. 1986 AWWA Distribution System Symp.* Denver, Colo.: American Water Works Association.

————. 1988a. Contaminant Propagation in Distribution Systems. *Jour. Environ. Eng.*, 114(2).

————. 1988b. Development, Application, and Calibration of Models for Predicting Water Quality in Distribution Systems. In *Proc. 1988 AWWA Water Quality and Treatment Conference*, Denver, Colo.: American Water Works Association.

————. 1988c. Modeling Contaminant Propagation in Drinking Water Distribution Systems. *Jour. Water Supply Research and Technology–Aqua*, 3:137–151.

Clark, R.M., W.M. Grayman, and R.M. Males. 1988. Contaminant Propagation in Distribution Systems. *Jour. Environ. Eng.*, 114(4).

Clark, R.M., and J.A. Coyle. 1990. Measuring and Modeling Variations in Distribution System Water Quality. *Jour. AWWA*, 82(8):46.

Clark, R.M., W.M. Grayman, J.A. Goodrich, R.A. Deininger, and A.F. Hess. 1990. Water Quality Modeling and Sampling Study in a Distribution System. In *Proc. 1990 AWWA Distribution System Symposium*. Denver, Colo.: American Water Works Association.

Cross, H. 1936. *Analysis of Flow in Networks of Conduits or Conductors*. Univ. of Ill. Eng. Experiment Station Bulletin 286.

Gessler, J., and T.M. Walski. 1985. *Water Distribution System Optimization*, TREL-85-11, WES, Vicksburg, Miss.: US Army Corps of Engineers.

Grayman, W.M., R.M. Clark, and R.M. Males. 1988a. Modeling Distribution-System Water Quality: Dynamic Approach. *Jour. Water Resources Planning and Management*, 114(3).

————. 1988b. A Set of Models to Predict Water Quality in Distribution Systems. In *Proc. Intern. Symp. on Computer Modeling of Water Dist. Systems*. Lexington, Ky.: University of Kentucky.

Hart, F.L., J.L. Meader, and S.N. Chiang. 1986. CLNET - A Simulation Model for Tracing Chlorine Residuals in a Potable Water Distribution Network. In *Proc. 1986 AWWA Distribution System Symposium*. Denver, Colo.: American Water Works Association.

Hart, F.L. 1991. Applications of the NET Software Package. In *Proc. AWWARF/EPA Conference on Water Quality Modeling in Distribution Systems*. Denver, Colo.: American Water Works Association Research Foundation and American Water Works Association.

Jeppson, R.W. 1976. *Analysis of Flow in Pipe Networks*. Ann Arbor, Mich.: Ann Arbor Science.

Kroon, J.R., and W.A. Hunt. 1989. Modeling Water Quality in the Distribution Network. In *Proc. 1989 AWWA Water Quality Technology Conference*. Denver, Colo.: American Water Works Association.

Kroon, J.R. 1990. An Innovation in Distribution System Water Quality Modeling. *Waterworld News*, 6(4).

Liou, C.P., and J.R. Kroon. 1986. Propagation and Distribution of Waterborne Substances in Networks. In *Proc. 1986 AWWA Distribution System Symposium*. Denver, Colo.: American Water Works Association.

Males, R.M., R.M. Clark, P.J. Wehrman, and W.E. Gates. 1985. Algorithm for Mixing Problems in Water Systems. *Jour. of Hyd. Eng.*, ASCE, 111(2).

Metzger, I. 1985. Water Quality Modeling of Distribution Systems. *ASCE Spec. Conf. on Comp. Applic. in Water Res.* New York: ASCE.

Sarikelle, S., Y. Chuang, and G.A. Loesch. 1989. Analysis of Water Distribution Systems on a Supercomputer. In *Proc. AWWA Comp. Spec. Conf.* Denver, Colo.: American Water Works Association.

Vasconcelos, J.J., P.F. Boulos, W.M. Grayman, L. Kiene, O. Wable, P. Biswas, A. Bhari, L.A. Rossman, R.M. Clark, and J.A. Goodrich. 1996. *Characterization and Modeling of Chlorine Decay in Distribution Systems*. Denver, Colo.: American Water Works Association Research Foundation and American Water Works Association.

Wood, D.J. 1980a. *Computer Analysis of Flow in Pipe Networks*. Lexington, Ky.: University of Kentucky.

———. 1980b. Slurry Flow in Pipe Networks. *Jour. Hydraulics*, 106(1).

Wood, D.J., and L.E. Ormsbee. 1989. Supply Identification for Water Distribution Systems. *Jour. AWWA*, 81(7):74.

Hydraulic Analysis

Throughout history people have been concerned with controlling floods, irrigation, and water supply. In the middle ages, pipelines were built sporadically in the Middle East to supply water from springs to nearby forts and palaces by gravity flow. More than 150 years ago, developments in iron and steel manufacturing and the introduction of efficient pumps allowed the building of pipelines to become common.

Despite its history, hydraulics is still somewhat empirical, with many indeterminate factors. However, enormous progress has been made in understanding the fundamental laws of fluid mechanics during the last century. Powerful mathematical techniques are available for solving these problems, but many practical hydraulics problems still do not have theoretical solutions. The amount of energy required to move water from one point to another and to provide the needed amount of water has become a major issue for designers. Since the amount of water moved is easily measurable, a great deal of experimental data has been collected on pipe flow. A consequence is that engineers and scientists have put these data into the form of graphs and empirical design formulas. Many of these equations are used in this chapter.

Friction Head Loss Formulas

A key factor in evaluating the flow through pipe networks is the ability to calculate friction head loss (Jeppson 1976). The three equations discussed in this chapter are the Darcy–Weisbach, the Hazen–Williams, and the Manning equations.

The Darcy–Weisbach equation is:

$$h_f = f(L/D)(V^2/2g) \qquad \text{(Eq 3-1)}$$

Where:

h_f = head loss, in ft/ft (m/m)

f = dimensionless fraction factor

D = pipe diameter, in ft (m)

L = length of pipe, in ft (m)

V = average velocity of flow, in ft/sec (m/sec)

g = the acceleration of gravity

A fundamental relationship that is important for hydraulic analysis is the Reynolds number, as follows:

$$R_e = VD/v \qquad \text{(Eq 3-2)}$$

Where:

R_e = Reynolds number (dimensionless)

v = kinametic viscosity, in ft^2/sec (m^2/sec)

and V and D are as defined earlier.

Table 3-1 contains various equations that can be used for calculating f in the Darcy–Weisbach equation. Although the Darcy–Weisbach equation is fundamentally sound, the most widely used equation is the Hazen–Williams equation

$$Q = 1.318CAR^{0.63}S^{0.54} \qquad \text{(Eq 3-3)}$$

Where:

Q = flow, in ft^3/sec (m^3/sec)

C = Hazen–Williams roughness coefficient

A = Cross-sectional area, in ft^2 (m^2)

R = Hydraulic radius (D/4) (m)

S = Slope of the energy grade line (h_f/L[ft/ft]) (m/m)

If the head loss is desired and Q is known, the Hazen–Williams equation for a pipe can be written as

$$h_f = (4.73L/C^{1.852}D^{4.87})Q^{1.8152} \qquad \text{(Eq 3-4)}$$

Where:

variables are as defined previously.

Another empirical equation is the Manning equation for pipes, which has been solved for h_f as follows:

$$h_f = (4.637 n^2 L / D^{5.333}) Q^2 \qquad \text{(Eq 3-5)}$$

Where:

n = empirical constant

and other variables are as defined earlier.

Table 3-2 contains values for the Hazen–Williams coefficient and the n values for various materials.

Table 3-1 Summary of friction factor equations for Darcy–Weisbach equation $h_f = f \dfrac{L}{D} \dfrac{V^2}{2g}$

Type of Flow	Equation Giving f	Range of Application
Laminar	$f = 64/R_e$	$R_e < 2{,}100$
Hydraulically smooth or turbulent smooth	$f = 0.316/R_e^{0.25}$	$4{,}000 < R_e < 10^5$
	$\dfrac{1}{\sqrt{f}} = 2 \log_{10}(R_e\sqrt{f}) - 0.8$	$R_e > 4{,}000$
Transition between hydraulically smooth and wholly rough	$\dfrac{1}{\sqrt{f}} = 2 \log_{10}\left(\dfrac{e/D}{3.7} + \dfrac{2.52}{R_e\sqrt{f}}\right)$	$R_e > 4{,}000$
	$= 1.14 - 2 \log_{10}\left(\dfrac{e}{D} + \dfrac{9.35}{R_e\sqrt{f}}\right)$	
Hydraulically rough or turbulent rough	$\dfrac{1}{\sqrt{f}} = 1.14 - 2 \log_{10}(e/D)$	$R_e > 4{,}000$
	$= 1.14 + 2 \log_{10}(D/e)$	

Where:

h_f = head loss, in ft/ft (m/m)

f = dimensionless fraction factor

L = length of pipe, in ft (m)

D = diameter of pipe, in ft (m)

V = average velocity of flow, in ft/sec (m/sec)

g = the acceleration of gravity

R_e = Reynolds number (dimensionless)

Analysis Methods

Analyzing for the flow in pipe networks, particularly if a large number of pipes are involved, is a complex process. Deciding which pipes should be included in the analysis can be a matter of judgment. It may not be possible or practical to include all the pipes that deliver water to the consumer. Analysis frequently is conducted only on the major transmission lines in the network or on the pipes that carry water between separate sections of the network. This process is called *skeletonization*.

As mentioned previously, there are two types of analyses usually conducted on drinking water distribution systems—steady state and dynamic—which were introduced in chapter 2.

The dynamic analysis or extended period simulation (EPS) deals with unsteady flows or transient problems. The steady state analysis, which will be discussed in this section, is considered solved when the flow rate in each pipe is calculated based on a specific usage or demand pattern and consumption. The supply from reservoirs, storage tanks, and/or pumps is generally the inflow or outflow from some point in the network. If flow rates are known, then pressures or head losses can be computed throughout the system. If the heads or pressures are known at each pipe junction or network node, the flow rates can be computed in each pipe.

A steady state analysis is needed for each demand or consumption pattern. The addition of new service areas, pumps, or storage tanks changes the system and requires a new steady state analysis. Dynamic analysis or EPS is often simply a series of steady state analyses linked by specified conditions. The oldest method of solving steady state flow in pipes is the Hardy Cross method. It was originally developed for solution by hand but has been programmed for solution using computers. However, when applied to large networks or for certain conditions it might be slow or even fail to converge. More recently the Newton–Raphson method and the "linear theory method," both described later in this chapter, have been applied to network solutions. The

Table 3-2 Values of the Hazen–Williams coefficient C_{HW} and Manning's n for common pipe materials

Type of Pipe	C_{HW}	n
Polyvinyl chloride (PVC) pipe	150	0.008
Very smooth pipe	140	0.011
New cast iron or welded steel	130	0.014
Wood or concrete	120	0.016
Clay or new riveted steel	120	0.017
Old cast iron or brick	100	0.020
Badly corroded cast iron or steel	80	0.035

Source: Jeppson (1976).

Newton–Raphson method requires approximately the same computer storage requirements as the Hardy Cross method and also requires an initial solution.

Other issues to be considered include the calibration and verification of the hydraulic assumptions. The issues of steady state, EPS, analysis, skeletonization, and calibration will also be discussed in this chapter.

Reducing Network Complexity

Pipe networks may include pipes in series, parallel, or branches (like branches of a tree). The network may also use elbows, valves, motors, and other devices that cause local disturbances and head losses. These factors can frequently be combined with or converted into equivalent pipes, which is very useful in simplifying networks. The major methods of simplification follow.

PIPES IN SERIES

For pipes in series, the same flow must pass through both pipes, therefore an equivalent head loss is the sum of the head losses for all the pipes being considered as part of the equivalent pipes. For example, Figure 3-1 depicts the head loss relationships for the pipes in series at different diameters.

The head loss for the equivalent pipe is

$$h_t = \Sigma h_i \qquad \text{(Eq 3-6)}$$

Where:

h_t = total head loss through both pipes

h_i = head loss through individual pipes

If we assume an exponential head loss relationship, such as in the Hazen–Williams formula, then

$$C_t Q^n = C_1 Q^{n_1} + C_2 Q^{n_2} + \dots \qquad \text{(Eq 3-7)}$$

Where:

C_t = empirical constant for head loss for entire length of pipe

Q = flow, in ft^3/sec (m^3/sec)

n = empirical constant

C_i = empirical constant for head loss for each length of pipe

If all of the ns are equal, then C_t equals the sum of the C_i for the individual pipes in series or

$$C_t = C_1 + C_2 + \dots \qquad \text{(Eq 3-8)}$$

PIPES IN PARALLEL

Two or more pipes in parallel can also be replaced by an equivalent pipe. In this case, the head loss between junctions where the pipes part and then join again must be equal. Although using equivalent pipes may be effective from a hydraulic viewpoint, this practice can be misleading when modeling water quality. Water quality modeling accuracy depends on accurate measurement of velocity.

Figure 3-2 illustrates the equivalent pipe approach.

$$h_t = h_1 + h_2 + \ldots + h_i \qquad \text{(Eq 3-9)}$$

The flow must equal the sum of the individual flows or

$$Q_t = Q_1 + Q_2 + \ldots + Q_i \qquad \text{(Eq 3-10)}$$

or

$$Q_t = \Sigma Q_i \qquad \text{(Eq 3-11)}$$

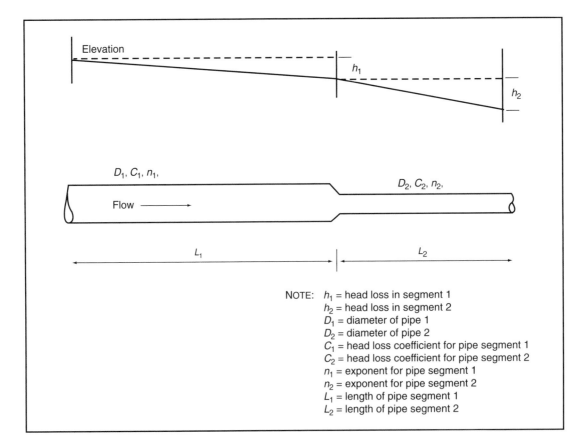

NOTE: h_1 = head loss in segment 1
h_2 = head loss in segment 2
D_1 = diameter of pipe 1
D_2 = diameter of pipe 2
C_1 = head loss coefficient for pipe segment 1
C_2 = head loss coefficient for pipe segment 2
n_1 = exponent for pipe segment 1
n_2 = exponent for pipe segment 2
L_1 = length of pipe segment 1
L_2 = length of pipe segment 2

Figure 3-1   Pipes of different diameters in series

Where:

Q_t =   total flow, in ft³/sec

Q_i =   flow, in ft³/sec in each parallel pipe

Using the exponential form for head loss yields

$$C_t Q_t^{n_t} = C_1 Q_1^{n_1} + C_1 Q_2^{n_2} + \dots + C_i Q_i^{n_i} \qquad \text{(Eq 3-12)}$$

$$C_t Q_t^{n_t} = \Sigma C_i Q_i^{n_i} \qquad \text{(Eq 3-13)}$$

Where:

$C_t Q_t^{n}$ =   sum of head loss in parallel pipes

$C_i Q_i^{n}$ =   head loss in each parallel pipe

If we assume the exponential form, then

$$\left(\frac{h_t}{C_t}\right)^{\frac{1}{n_t}} = \left(\frac{h_1}{C_1}\right)^{\frac{1}{n_1}} + \left(\frac{h_2}{C_2}\right)^{\frac{1}{n_2}} + \dots \qquad \text{(Eq 3-14)}$$

Since Eq 3-9 holds and assuming the exponents are equal, then

$$\left(\frac{1}{C_t}\right)^{\frac{1}{n}} = \left(\frac{1}{C_1}\right)^{\frac{1}{n}} + \left(\frac{1}{C_2}\right)^{\frac{1}{n}} + \dots \qquad \text{(Eq 3-15)}$$

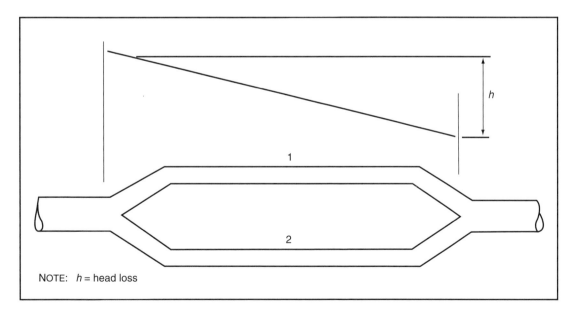

NOTE: h = head loss

Figure 3-2   Head loss over a set of parallel pipes

$$\left(\frac{1}{C_t}\right)^{\frac{1}{n}} = \Sigma_i\left(\frac{1}{C_i}\right)^{\frac{1}{n}}$$
(Eq 3-16)

If the Darcy–Weisbach equation is used, it is common to assume n is equal for all pipes and Eq 3-16 can be used to solve for the C_t in the equivalent pipe.

BRANCHING SYSTEM

In a branching system, a number of pipes are connected to a larger pipe in the form of a tree (Figure 3-3). If the flow is from the larger pipe to the smaller laterals, the flow rate can be calculated in any pipe as the sum of the downstream consumption or demand. If the laterals supply water to the main, then the same approach can be taken. When a system is analyzed, frequently only the larger pipes are used in the network analysis.

MINOR LOSSES

An equivalent pipe can be used in a network to approximate the minor losses associated with valves, meters, elbows, or other devices. Equivalent pipes are formed by adding a length to the actual pipe length that will result in the same head loss as in the component. The Darcy–Weisbach or Hazen–Williams equations can be used to compute the head loss.

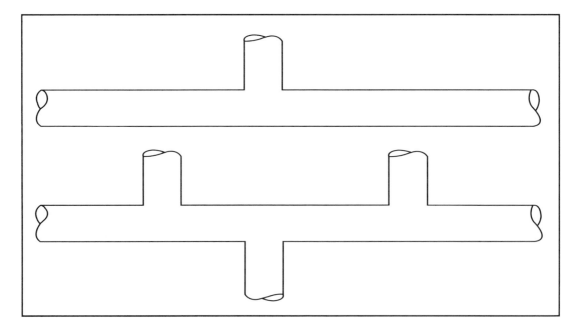

Figure 3-3 Typical branching pipes

Equations Describing Flow

Flow analysis in pipe networks is based on basic continuity and energy laws. The mass weight or volumetric flow rate into a junction must equal the mass weight or volumetric flow rate out of a junction, including demand. This relationship can be expressed as follows:

$$\Sigma Q_{out} - \Sigma Q_{in} = D \qquad \text{(Eq 3-17)}$$

Where:

Q_{out} = flow away from junction

Q_{in} = flow into junction

D = demand at the junction

In addition to continuity equations that must be satisfied, the head loss or energy equations must be satisfied. If one sums the head loss around a loop, the net head loss must equal zero. This relationship gives N equations of the form

$$\sum_{i} h_i = 0$$

$$\sum_{i}^{2} h_i = 0$$

$$\vdots$$

$$\sum^{N} h_i = 0 \qquad \text{(Eq 3-18)}$$

Where:

N = the number of nonoverlapping loops in the network, and the summation on i is over the pipes in the loops 1, 2...N.

Using the exponential formula $h_i = C_i Q_i^{n_i}$ can be written in terms of the flow rate, or

$$\sum_i C_i Q_i^{n_i} = 0$$

$$\sum_i^2 C_i Q_i^{n_i} = 0 \qquad \text{(Eq 3-19)}$$

$$\vdots$$

$$\sum_i^N C_i Q_i^{n_i} = 0$$

The following equation describes a pipe network with junctions J, non-overlapping loops N, and pipes P

$$P = (J - 1) + N \qquad \text{(Eq 3-20)}$$

The flow rate in each pipe can be considered unknown, resulting in P unknowns. The number of independent equations that can be obtained for a network described above are $(J - 1) + N$. Therefore, the number of independent equations is equal in number to the unknown flow rates in the P pipes. The $(J - 1)$ continuity equations are linear and N energy (or head losses) equations are nonlinear. Many networks consist of hundreds of pipes, therefore systematic networks that use computers are needed for solving these systems of simultaneous equations.

Figure 3-4 is a simple network with one demand and one source. The equations that describe energy loss are shown below:

$$h_{12} = C_{12} Q_{12}^n$$
$$h_{23} = C_{23} Q_{23}^n$$
$$h_{45} = C_{45} Q_{45}^n$$
$$h_{56} = C_{56} Q_{56}^n \qquad \text{(Eq 3-21)}$$
$$h_{14} = C_{14} Q_{14}^n$$
$$h_{25} = C_{25} Q_{25}^n$$
$$h_{36} = C_{36} Q_{36}^n$$

Where:

h_{ij} = the head loss in link ij in the Hazen–Williams equation

C_{ij} = the coefficient for the pipe ij

$$Q_{ij} = \text{the flow in pipe link ij}$$

$$n = \text{the exponent in the Hazen–Williams equation}$$

The conservation of mass equations are

$$Q = Q_{12} + Q_{14}$$
$$Q_{12} = Q_{25} + Q_{23}$$
$$Q_{23} = Q_{36}$$
$$Q_{14} = Q_{45}$$
$$Q_{56} = Q_{45} + Q_{25}$$
$$Q = Q_{56} + Q_{36}$$

(Eq 3-22)

Between Eq 3-21 and Eq 3-22 there are 13 equations and 13 unknowns (6 unknown heads and 7 unknown flow rates).

HEADS AT JUNCTIONS AS UNKNOWNS

If the head (either the total head or the piezometric head, since the velocity head is generally ignored in determining heads or pressure in pipe networks) at each junction is initially considered unknown instead of the flow rate in each pipe, the number of simultaneous equations that must be solved can be reduced in number (Jeppson 1976). The reductions in number of equations, however, is at the expense of not having some linear equations in the system.

To obtain the system of equations that contains the heads at the junctions of the network as unknowns, the $J - 1$ independent continuity equations are

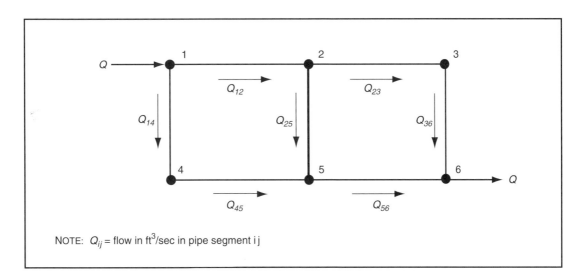

NOTE: Q_{ij} = flow in ft^3/sec in pipe segment i j

Figure 3-4 Steady state network hydraulics

written as before. Thereafter, the relationship between the flow rate and head loss is substituted into the continuity equations. In writing these equations, it is convenient to use a double subscript for the flow rates. These subscripts correspond to the junctions at the ends of the pipe. The first subscript is the junction number from which the flow comes. The second number is the junction number to which the flow goes. Thus Q_{12} represents the flow in the pipe connecting junctions 1 and 2, assuming the flow is from junction 1 to junction 2. If the flow is actually in this direction, Q_{12} is positive and Q_{21} equals minus Q_{12}. Solving for Q from the exponential formula (using the double subscript notation) gives

$$Q_{ij} = \left(\frac{h_{ij}}{C_{ij}}\right)^{\frac{1}{n}}$$

(Eq 3-23)

$$Q_{ij} = \left(\frac{(h_i - h_j)}{C_{ij}}\right)^{\frac{1}{n}}$$

Substituting Eq 3-23 into Eq 3-17 yields

$$C = \left[\sum\left(\frac{(h_i - h_j)}{C_{ij}}\right)^{\frac{1}{n}}\right]_{out} - \left[\sum\left(\frac{(h_i - h_j)}{C_{ij}}\right)^{\frac{1}{n}}\right]_{in}$$

(Eq 3-24)

Equations of the form of Eq 3-24 are written at $J - 1$ junctions, producing a system of $J - 1$ nonlinear equations. Upon solving this nonlinear system of equations, the pressure at any junction j can be computed by subtracting the junction elevation from the head h_i and then multiplying this difference by the specific weight of water. To determine the flow rates in the pipes of the network, known heads are substituted into Eq 3-23.

CORRECTIVE FLOW RATES AROUND LOOPS OF NETWORK CONSIDERED UNKNOWNS

Since the number of junctions minus 1 (i.e., $J - 1$) will be less than the number of pipes in a network by the number of loops N in the network, the last set of head loss equations will generally be less in number than the system of mass balance equations (Jeppson 1976). The head loss equations will be nonlinear, but the flow equations will be linear with the exception of the energy equation, which will be nonlinear. Another approach is to write a series of equations that include a corrective flow rate in each loop as an unknown. This set of equations will be referred to as the ΔQ-equations. Since there are N basic loops in a network, the ΔQ-equations consist of N equations, all of which are nonlinear.

It is not difficult to establish an initial flow in each pipe that satisfies the continuity equations at each junction (which must also satisfy the jth junction continuity equation). These initial flow estimates generally will not simultaneously satisfy the N head loss equations and must, therefore, be corrected before they equal the true flow rates in the pipes. A flow rate adjustment or ΔQ can be added (accounting for sign) to the initially assumed flow in each pipe forming a loop of the network without violating continuity conditions at the junctions. This fact suggests establishing N energy (or head loss) equations around the N loops of the network in which the initial flow plus the correction rate ΔQ is used as the true flow rate in the head loss equations. Once these head loss equations are solved, the $J - 1$ continuity equations would remain satisfied as they initially were. The corrective loop flow rates ΔQ_i may be taken as positive or negative around the basic loops, but the sign convention must be consistent around any particular loop, and preferably in the same direction of all loops of the network. For purposes of this analysis, clockwise directions will be considered positive.

Let's assume an initial flow in a pipe of Q_{ij} (assume P pipes) and that the corrective loop flow rates will be denoted by ΔQ_{ij}. The energy equation for Figure 3-4 would be written as follows:

$$C_{12}(Q_{12} + \Delta Q_{12})^{n_{12}} + C_{25}(Q_{25} + \Delta Q_{25})^{n_{25}}$$
$$+ C_{54}(Q_{54} + \Delta Q_{54})^{n_{54}} + C_{41}(Q_{41} + \Delta Q_{41})^{n_{41}} = 0 \qquad \text{(Eq 3-25)}$$

$$C_{23}(Q_{23} + \Delta Q_{23})^{n_{23}} + C_{36}(Q_{36} + \Delta Q_{36})^{n_{36}}$$
$$+ C_{65}(Q_{65} + \Delta Q_{65})^{n_{65}} + C_{52}(Q_{52} + \Delta Q_{52})^{n_{52}} = 0$$

The first step in solving these equations is to provide initial estimates of the flow rate in each pipe that satisfy the continuity equations at the junctions and then solving for ΔQ_{ij}. Once the solution to ΔQ_{ij} is found, the flow rate for each pipe can be quickly determined. There are several common techniques for solving the sets of equations discussed in this section. These include the linear method and the Newton–Raphson method, which will be discussed in the following sections.

Solution Techniques

LINEAR THEORY

Linear theory transforms the N nonlinear loop equations into linear equations by approximating the head in each pipe by

$$h_{ij} = \{[C_{ij}Q_{ij}(0)]^{n-1}\}Q_{ij} \qquad \text{(Eq 3-26)}$$

$$h_{ij} = C_{ij}'Q_{ij}$$

Where:

h_{ij} = the head loss in link ij in the Hazen–Williams equation

C_{ij} = the coefficient for the pipe ij

Q_{ij} = the flow in pipe link ij

n = the exponent in the Hazen–Williams equation

A coefficient is defined for each pipe as the product of C_{ij} multiplied by $[Q_{ij}(0)]^{n-1}$, an estimate of the flow rate in that pipe (Jeppson 1976). Combining the equation set represented by Eq 3-26 with the $J-1$ junction continuity equations provides a system of L linear equations that can be solved by linear algebra. After a series of iterations, the $Q_{ij}(m)s$ will equal the $Q_{ij}s$. Pumps, valves, and pressure-reduction valves can be incorporated with the linear theory method.

NEWTON–RAPHSON METHOD

The Newton–Raphson technique is widely applied and can be used to solve any of the following types of equations describing flow in pipes: the flow rate in each pipe, the head at each junction, and the correct flow rate around each loop. In applying the Newton–Raphson method, a solution to the equation $F(x) = 0$ is obtained by applying the iterative formula $x^{m+1} = x^m - F(x^m)/F^l(x^m)$ (Jeppson 1976).

Network Characterization

Engineering analysis of water distribution systems is frequently limited to the solution of the hydraulic network problem, i.e., given the physical characteristics of a distribution system modeled as a node–link network and the demands at nodes (junctions where network components connect to one another), the flows in links (a connection of any two nodes) and head at all nodes of the network are determined. This problem is formulated as a set of simultaneous nonlinear equations, and a number of well-known solution methods exist, many of which have been coded as computer programs known as *hydraulic network models* (Cesario 1995).

Mathematical methods for analyzing the flow and pressure in networks have been in use for more than 50 years. They are generally based on well-accepted hydrodynamic equations. Computer-based models for performing this type of analysis were first developed in the 1950s and 1960s and greatly expanded and made more available in the 1970s and 1980s. Currently, dozens of such models are readily available on systems ranging from personal computers to supercomputers.

Hydraulic models were developed to simulate flow and pressures in a distribution system either under steady state conditions or under time varying demand (extended period simulation) and operational conditions. Hydraulic models may also incorporate optimization components that aid the user in selecting system parameters that result in the best match between observed system performance and model results.

APPLICATION OF MODELS

The following steps should be followed in applying network models (Clark et al. 1988):

1. *Model selection.* Define model requirements and select a model (hydraulic and/or water quality) that fits your requirements, style, budget, etc.

2. *Network representation.* Accurately represent the distribution system components in the model.

3. *Calibration.* Adjust model parameters so that predicted results adequately reflect observed field data.

4. *Verification.* Independently compare model and field results to verify the adequacy of the model representation.

5. *Problem definition.* Define the specific design or operational problem to be studied and incorporated (i.e., demands and system operation) into the model.

6. *Model application.* Use the model to study the specific problem or situation.

7. *Display and analysis of results.* Following the model application, display and analyze the results to determine how reasonable the results are. Finally, translate the results into a solution to the problem.

As mentioned in previous chapters, all water distribution system models represent the network as a series of connected links and nodes. Links are defined by the two end nodes and generally serve as conveyance devices (e.g., pipes), while nodes represent point components (e.g., junctions, tanks, and treatment plants). In most models, pumps and valves are represented as links. Water demands are aggregated and assigned to nodes; an obvious simplification of real-world situations in which individual house taps are distributed along a pipe rather than at junction nodes.

Network representation is more of an art than a science. Most networks are skeletonized for analysis, which means that they contain only a representation of selected pipes in a service area.

A minimal skeletonization should include all pipes and features of major concern. Nodes are usually placed at pipe junctions, where pipe characteristics may change in diameter, *C*-value (roughness), or material of construction.

Nodes may also be placed at locations of known pressure or where pressure valves are desired.

Network components include pipes, reservoirs, pumps, pressure-relief valves, control valves, altitude valves, check valves, and pressure-reducing valves. For steady state modeling, a key input is information on water consumption. Usage is assigned to nodes in most models and may be estimated several different ways. Demand may be estimated by a count of structures of different types using a representative consumption per structure, meter readings and the assignment of each meter to a node, and to general land use. A universal adjustment factor should be used so that total usage in the model corresponds to total production. Figures 3-5 and 3-6 illustrate two methods of estimating water use.

MODELING TEMPORAL VARIATIONS

We know that most phenomena and behavior vary over time in the real world. Models that assume no variation over time are referred to as steady state models, which provide a snapshot at a single point in time. In distribution system models, steady state models assume that demands are constant and operations are constant (e.g., tank water-level elevations remain constant and pumps are either on or off at a constant speed). Though such assumptions may not be valid over long periods of time, steady state models can provide some useful information concerning the behavior of a network under various representative conditions, including fire demand, nighttime low demands, etc.

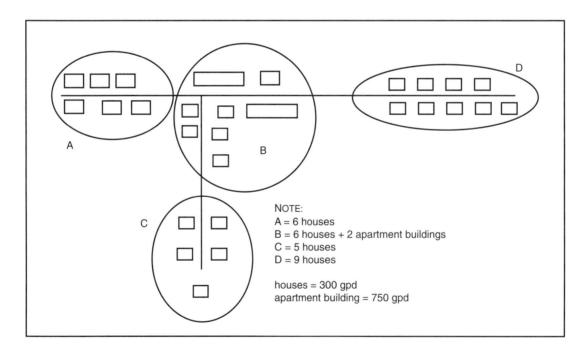

Figure 3-5 Usage by counting buildings

Models that allow for variations over time are referred to as *temporally dynamic models*. Most network distribution hydraulic models incorporate temporal variation by stringing together a series of steady state solutions and refer to this method as extended period simulation (EPS). For example, in the EPS mode, demand at a node may be assumed to be 100 gpm (378 L/min) from 2:00 p.m. to 3:00 p.m. and then 150 gpm (567 L/min) from 3:00 p.m. to 4:00 p.m. Thus, one steady state solution may be assumed from 2:00 p.m. until 3:00 p.m. and a different solution may be calculated from 3:00 p.m. to 4:00 p.m. Further temporal dynamics may be incorporated by checking the water level in a tank and, if the water level reaches a maximum allowable level at 3:20 p.m., then another steady state solution is started at 3:20 p.m. with the tank discharging instead of filling. Though the EPS solution does introduce some approximations and totally ignores the transient phenomena resulting from sudden changes, such as a pump being turned on, these more refined assumptions are generally not considered significant for most distribution system studies. Figure 3-7 illustrates systemwide demand variation.

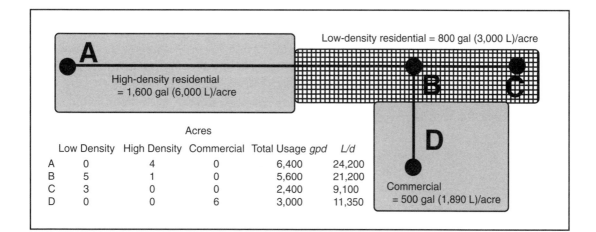

Figure 3-6   Water usage by land use

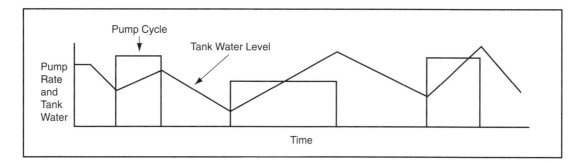

Figure 3-7   Systemwide temporal analysis

Some of the characteristics of the EPS mode follow:

- requires information on temporal variations in water usage over the period being modeled

- permits temporal patterns to be defined for groups of nodes (spatially different patterns can be applied to a given node)

- uses the best available information on temporal patterns that can be estimated (for example, some users have continuous meters)

- can sometimes use literature values for a first guess at residential patterns (which may vary with climate)

- can use analysis of information from Supervisory Control and Data Aquisition (SCADA) to estimate systemwide temporal pattern

MODEL CALIBRATION

Calibration is an important part of the "art" of modeling water distribution systems. Model calibration is the process of adjusting model input data (or, in some cases, model structure) so that the simulated hydraulic and water quality output sufficiently mirrors observed field data. As mentioned earlier, one way of viewing calibration is to think of a TV screen showing observed and predicted values. Calibration is the process of adjusting the picture so that the predicted value provides the closest estimate for the actual values.

Depending on the degree of accuracy, calibration can be difficult, costly, and time-consuming. The extent and difficulty of calibration is minimized by developing an accurate representation of the network and its components. A traditional technique for calibration is to use fire flow pressure measurements. Pressures and flow in isolated pipe sections are measured in the field and the roughness factors C are adjusted to reflect the data.

Another method is to use water quality tracers. Naturally occurring or added chemical tracers may be measured in the field and the results used to calibrate hydraulic and water quality models. The most common tracer is fluoride. It is conservative (does not degrade), safe, and can usually be added (or normal feed can be curtailed) and the movement can be traced in the system using hand-held analyzers. For conservative tracers, adjustments may be made primarily in the hydraulic model to adequately match the predicted and observed concentrations.

Another calibration technique is to measure predicted tank levels derived from computer simulations against actual tank level during a given period of record. For example, using data from SCADA systems or from on-line pressure and tank-level recorders, flows can be adjusted in the simulation model until they match the actual tank-level information.

References

Cesario, L. 1995. *Modeling Analysis and Design of Water Distribution Systems.* Denver, Colo.: American Water Works Association.

Clark, R.M., W.M. Grayman, R.M. Males, and J. Coyle. 1988. Modeling Contaminant Propagation in Drinking Water Distribution Systems. *Aqua,* 37(3):137–151.

Jeppson, R.W. 1976. *Analysis of Flow in Pipe Networks.* Ann Arbor, Mich.: Ann Arbor Science.

Water Quality Models

Background

In order to model water quality within distribution systems, the concentration of a particular substance must be calculated as it moves through the system from various points of entry (e.g., treatment plants) and on to water users. This movement is based on three principles

1. Conservation of mass within differential lengths of pipe.

2. Complete and instantaneous mixing of the water entering pipe junctions.

3. Appropriate kinetic expressions for the growth or decay of the substance as it flows through pipes and storage facilities.

This change in concentration can be expressed by the following differential equation:

$$\frac{\partial C_{ij}}{\partial t} = -v_{ij}\frac{\partial C_{ij}}{\partial x} + k_{ij}C_{ij} \qquad \text{(Eq 4-1)}$$

Where:

C_{ij} = substance concentration (g/m^3) at position x and time t in the link between nodes i and j

v_{ij} = flow velocity in the link (equal to the link's flow divided by its cross-sectional area) (m/sec)

k_{ij} = rate at which the substance reacts within the link (sec^{-1})

This equation shows that the rate at which the mass of material changes within a small section of pipe equals the difference in mass flow into and out of the section plus the rate of reaction within the section. It is assumed that the velocities in the links are known beforehand from the solution to a hydraulic model of the network. In order to solve Eq 4-1 we need to know C_{ij} at $x = 0$ for all times (a boundary condition) and a value for k_{ij}.

Equation 4-2 represents the concentration of material leaving the junction and entering a pipe

$$C_{ij}\big|_{x = 0} = \frac{\sum_{k} Q_{ki} C_{kj}\big|_{x = L}}{\sum_{L} Q_{kj}} \qquad \text{(Eq 4-2)}$$

Where:

$C_{ij}\big|_{x = 0}$ = the concentration at the start of the link connecting node i to node j in mg/L (i.e., where $x = 0$)

$C_{kj}\big|_{x = L}$ = the concentration at the end of a link, in mg/L

Q_{ki} = flow from k to i

Equation 4-2 implies that the concentration leaving a junction equals the total mass of a substance mass flowing into the junction divided by the total flow into the junction.

Storage tanks can be modeled as completely mixed, variable volume reactors in which the change in volume and concentration over time are as follows:

$$\frac{dV_s}{dt} = \sum_{k} Q_{ks} - \sum_{i} Q_{sj} \qquad \text{(Eq 4-3)}$$

$$\frac{d(V_s C_s)}{dt} = \sum_{k} Q_{ks} C_{ks}\big|_{x = L} - \sum_{i} Q_{sj} C_s + k_{ij}(C_s) \qquad \text{(Eq 4-4)}$$

Where:

C_s = the concentration for tank s, in mg/L

dt = change in time, in seconds

Q_{ks} = flow from node k to s, in ft^3/sec (m^3/sec)

Q_{sj} = flow from node s to j, in ft^3/sec (m^3/sec)

dV_s = change in volume of tank at node s, in ft^3 (m^3)

V_s = volume of tank at node s, in ft^3 (m^3)

C_{ks} = concentration of contaminant at end of links, in mg/ft^3 (mg/m^3)

k_{ij} = decay coefficient between node i and j, in sec^{-1}

Previous chapters have described the historical development and application of water quality contaminant propagation models. This chapter describes some of the models and approaches being used in modeling contaminant propagation in drinking water distribution systems. Most of the discussion will be focused on a US Environmental Protection Agency (USEPA) developed hydraulic/contaminant propagation model called EPANET (Rossman 1994), which is based on mass transfer concepts (transfer of a substance through another on a molecular scale). Another approach to water quality contaminant propagation developed by Biswas, Lu, and Clark (1993) uses a steady state transport equation that takes into account the simultaneous corrective transport of chlorine in the axial direction, diffusion in the radial direction, and consumption by first- order reaction in the bulk liquid phase. Islam (1995) developed a model called QUALNET, which predicts the temporal and spatial distribution of chlorine in a pipe network under slowly varying unsteady flow conditions. Boulos, Altman, Jarrige, and Collevati (1995) proposed a technique called the Event Driven Method (EDM), which is based on a "next-event" scheduling approach, which can significantly reduce computing times.

Solution Methods

There are several different numerical methods that can be used to solve contaminant propagation equations. Four commonly used techniques are Eulerian finite-difference method (FDM), Eulerian discrete volume method (DVM), Lagrangian time-driven method (TDM), and Lagrangian event-driven method (EDM).

The FDM approximates derivatives with finite-difference equivalents along a fixed grid of points in time and space. Islam used this technique to model chlorine decay in distribution systems. This technique will be discussed in more detail later in this chapter (Islam 1995).

The DVM divides each pipe into a series of equally sized, completely mixed volume segments. At the end of each successive water quality time step, the concentration within each volume segment is first reacted and then transferred to the adjacent downstream segment. This approach was used in the models that were the basis for the early USEPA studies.

The TDM tracks the concentration and size of a nonoverlapping segment of water that fills each link of a network. As time progresses, the size of the most upstream segment in a link increases as water enters the link. An equal loss in size of the most downstream segment occurs as water leaves the link. The size of these segments remains unchanged.

The EDM is similar to TDM, except that rather than update an entire network at fixed time steps, individual link–node conditions are updated only at times when the leading segment in a link completely disappears through this downstream node.

EPANET

As mentioned previously, the EPANET hydraulic model has been a key component in providing the basis for water quality modeling. There are many commercially available hydraulic models that also incorporate water quality models as well. EPANET is a computer program based on the EPS approach to solving hydraulic behavior of a network. In addition, it is designed to be a research tool for modeling the movement and fate of drinking water constituents within distribution systems. EPANET calculates all flows in cubic feet per second (ft^3/s) and has an option for accepting flow units in gallons per minute (gpm), million gallons per day (mgd), or litres per second (L/s). Instructions on accessing EPANET through the Internet are included in chapter 10.

EPANET uses the Hazen–Williams formula, the Darcy–Weisbach formula, or the Chezy–Manning formula for calculating the head loss in pipes. Pumps, valves, and minor loss calculations in EPANET are also consistent with the convention established in the previous chapters. All nodes have their elevations above sea level specified and tanks and reservoirs are assumed to have a free water surface. The hydraulic head is simply the elevation of the surface above sea level. The surface of tanks is assumed to change in accordance with the following equation:

$$\Delta y = (q/A) \Delta t \qquad \text{(Eq 4-5)}$$

Where:

Δy = change in water level, in ft (m)

q = flow rate into (+) or out of (–) tank, in ft^3/s (m^3/s)

A = cross-sectional area of the tank, in ft (m)

Δt = time interval, in sec

It is assumed that water usage rates, external water supply rates, and source concentrations at nodes remain constant over a fixed period of time, although these quantities can change from one period to another. Nodes are junctions where network components connect to one another (links), as well as tanks and reservoirs. The default period interval is one hour but can be set to any desired value. Various consumption or water usage patterns can be assigned to individual nodes or groups of nodes.

EPANET solves a series of equations for each link using the gradient algorithm. Gradient algorithms provide an interactive mechanism for approaching an "optimal solution" by calculating a series of slopes that lead to "better and better" solutions. Flow continuity is maintained at all nodes after the first iteration. The method easily handles pumps and valves.

The set of equations solved for each link (between nodes i and j) and each node k is as follows:

$$h_i - h_j = f(q_{ij}) \qquad \text{(Eq 4-6)}$$

$$\Sigma_i\, q_{ik} - \Sigma_j\, q_{kj} - Q_k = 0 \qquad \text{(Eq 4-7)}$$

Where:

q_{ij} = flow in link connecting nodes i and j, in ft³/s (m³/s)

h_i = hydraulic grade line elevation at node i (equal to elevation head plus pressure head), in ft (m)

Q_k = flow consumed (+) or supplied (−) at node k, in ft³/s (m³/s)

$f(\cdot)$ = functional relation between head loss and flow in a link

The set of equations for each storage node (tank or reservoir) in the system is as follows:

$$\frac{\partial y_s}{\partial t} = \frac{q_s}{A_s} \qquad \text{(Eq 4-8)}$$

$$q_s = \Sigma_i q_{is} - \Sigma_j q_{sj} \qquad \text{(Eq 4-9)}$$

$$h_s = E_s + y_s \qquad \text{(Eq 4-10)}$$

Where:

y_s = height of water stored at node s, in ft (m)

A_s = cross-sectional area of storage node s (infinite for reservoirs), in ft² (m²)

E_s = elevation of node s, in ft (m)

q_s = flow into storage node s, in ft (m)

t = time, in seconds

h_s = height of water in storage tank, in ft (m)

Equation 4-8 expresses conservation of water volume at a storage node. Equations 4-9 and 4-10 express the same relationship for pipe junctions and Eq 4-6 represents energy loss or gain due to flow within a link. For known initial storage node levels y_s at time zero, Eq 4-6 and Eq 4-7 are solved for all flows q_{ij} and heads h_i using Eq 4-10 as a boundary condition. This is called *hydraulically balancing* the network and is accomplished by using an iterative technique to solve the resulting nonlinear equations.

After a network hydraulic solution is obtained, flow into (or out of) each storage node q_s is found using Eq 4-9 and used in Eq 4-8 to find new storage elevations after a time step dt. This process is then repeated for all following time steps for the remainder of the simulation period.

WATER QUALITY SIMULATION

EPANET uses the flows from the hydraulic simulation to track the propagation of contaminants through a distribution system. A conservation of mass equation is solved for the substance within each link between nodes i and j as follows:

$$\frac{\partial c_{ij}}{\partial t} = \frac{q_{ij}}{A_{ij}}\left(\frac{\partial c_{ij}}{\partial x_{ij}}\right) + \Theta(c_{ij}) \qquad \text{(Eq 4-11)}$$

Where:

c_{ij} = concentration of substance in link i, j as a function of distance and time (i.e., $C_{ij} = C_{ij}[X_{ij},t]$), in mass/ft^3 (mass/m^3)

x_{ij} = distance along link i, j, in ft (m)

q_{ij} = flow rate in link i, j at time t, in ft^3/s (m^3/s)

a_{ij} = cross-sectional area of link i, j, in ft^2 (m^2)

$\Theta(c_{ij})$ = rate of reaction of constituent within link i, j, in mass/ft^3/d (mass/m^3/d)

Equation 4-11 must be solved with known initial conditions at time zero. The following boundary condition at the beginning of the link, i.e., at node i where $x_{ij} = 0$ must hold

$$c_{ij}(0, t) = \frac{\Sigma_k q_{ki} c_{ki}(L_{ki},t) + M_i}{\Sigma_k q_{ki} + Q_{si}} \qquad \text{(Eq 4-12)}$$

The summations are made over all links k, i that flow into the head node i of the link i, j,

Where:

L_{ki} = the length of link k, i

M_i = the substance mass introduced by any external source at node i

Q_{si} = the source's flow rate

The boundary condition for link k, i depends on the concentrations at the head of the nodes of all links k, i that flow into link i, j. Equations 4-11 and 4-12 form a coupled set of differential/algebraic equations over all links in the network.

These equations are solved within EPANET by using DVM, which was described earlier.

Water quality time steps are chosen to be as large as possible without causing the flow volume of any pipe to exceed its physical volume. Therefore the water quality time step dt_{wq} source cannot be larger then the shortest time of travel through any pipe in the network, i.e.,

$$dt_{wq} = Min\left(\frac{V_{ij}}{q_{ij}}\right) \text{ for all pipes i, j} \qquad \text{(Eq 4-13)}$$

Where:

V_{ij} = the volume of pipe i, j

q_{ij} = the flow rate of pipe i, j

Pumps and valves are not part of this determination because transport through them is assumed to occur instantaneously. Based on this water quality time step, the number of volume segments in each pipe (n_{ij}) is

$$n_{ij} = INT\left(\frac{V_{ij}}{q_{ij}dt_{wq}}\right) \qquad \text{(Eq 4-14)}$$

Where:

$INT(x)$ = the largest integer less than or equal to x. There is both a default limit of 100 for pipe segments or the user can set dt_{wq} to be no smaller than a user-adjustable time tolerance.

REACTION RATE MODEL

Equation 4-1 provides a mechanism for evaluating the reaction of a substance as it travels through a distribution system (Rossman, Clark, and Grayman, 1994). Reaction can occur in the bulk phase or with the pipe wall. EPANET models both types of reactions using first-order kinetics. In general, within any given pipe, material in the bulk phase and at the pipe well will decrease according to the following equation:

$$\Theta(c) = -k_b c - \left(\frac{k_f}{R_H}\right)(c - c_w) \qquad \text{(Eq 4-15)}$$

Where:

$\Theta(c)$ = total reaction rate

k_b = first-order bulk reaction rate, in constant/sec

c = substance concentration in bulk flow, in mass/ft^3 (mass/m^3)

k_f = mass transfer coefficient between bulk flow and pipe wall, in ft/sec (m/sec)

R_H = hydraulic radius of pipe (pipe radius/2), in ft (m)

c_w = substance concentration at the wall, in mass/ft^3 (mass/m^3)

Assuming a mass balance for the substance at the pipe wall yields:

$$k_f(c - c_w) = k_w - c_w \qquad \text{(Eq 4-16)}$$

Where:

k_w = rate of chlorine demand at wall (wall demand), in ft/sec (m/sec)

i = an overall first order reaction rate is equal to:

$$K = k_b + (k_w k_f) / R_H (k_w + k_f) \qquad \text{(Eq 4-17)}$$

Where:

K = the overall reaction rate coefficient

Equation 4-16 pertains to the growth or decay of a substance, with mass transfer to or from the pipe wall depending on whether the sign of the equation is positive or negative. A negative sign means decay and a positive sign means growth.

There are three coefficients used by EPANET to describe reactions within a pipe. These are the bulk rate constant k_b and its wall rate constant k_w, which must be determined empirically and supplied as input to the model. The mass transfer coefficient is calculated internally by EPANET.

OTHER FEATURES

EPANET can also model the changes in age of water and travel time to a node. The percentage of water reaching any node from any other node can also be calculated. Source tracing is a useful tool for showing the degree that water from a given source blends with that from other sources, and how this pattern changes over time.

INPUT AND OUTPUTS

The following input steps should be taken before using EPANET:

1. Identify all network components and their connections. These components include pipes, pumps, valves, storage tanks, and reservoirs.

2. Assign unique ID numbers to all nodes. These numbers must be between 1 and 2,147,483,647, but do not have to be in any specific order.

3. Assign ID numbers to each link (pipe, pump, or valve). Both a link and a node can have the same ID number.

4. Collect information on the following system parameters:

    a. Diameter, length, roughness, and minor loss coefficient for each pipe.
    b. Characteristic operating curve for each pump.
    c. Diameter, minor loss coefficient, and pressure or flow setting for each control valve.
    d. Diameter and lower and upper water levels for each tank.
    e. Control rules that determine how pump, valve, and pipe settings change with time, tank water levels, or nodal pressures.
    f. Changes in water demands for each node over the time period being simulated.

g. Initial water quality at all nodes and changes in water quality over time at source nodes.

Output from EPANET includes

· color-coded network maps

· time series plots

· tabular reports

Table 4-1 contains a summary of input parameter units for EPANET.

Table 4-1   Summary of input parameter units

Parameter	US Customary Units	SI (Metric) Units
Junction elevation	feet	metres
Junction demand	flow units	flow units
Tank bottom elevation	feet	metres
Tank levels	feet above bottom	metres above bottom
Tank diameter	feet	metres
Tank minimum volume	cubic feet	cubic metres
Junction/tank quality		
chemical	milligrams/litre (or user supplied)	same as US customary
age	hours	same as US customary
source trace	percent	same as US customary
Pipe length	feet	metres
Pipe diameter	inches	millimetres
Pipe roughness		
Hazen–Williams	none	none
Darcy–Weisbach	millifeet	millimetres
Chezy–Manning	none	none
Minor loss coefficient	none	none
Pump power rating	horsepower	kilowatts
Pump head	feet	metres
Pump flow	flow units	flow units
Pump speed setting	none	none
Valve diameter	inches	millimetres
Valve pressure setting	pounds per square inch	metres
Valve flow setting	flow units	flow units
Bulk reaction coefficient	days−1	days−1
Wall reaction coefficient	feet/day	metres/day
Specific gravity	none	none
Viscosity	square feet per second	square metres per second
Diffusivity	square feet per second	square metres per second

Convective Transport Model

Biswas, Lu, and Clark (1993) developed a steady state transport equation that takes into account the simultaneous convective transport of chlorine in the axial direction, diffusion in the radial direction, and consumption by first order in the bulk liquid phase. Different wall conditions are considered in the model, including a perfect sink, no wall consumption, and partial wall consumption. The transport equations are developed in terms of the dissolved chlorine concentrations $C*(HOCl^+ + OCl^-)$ in mg/L^{-1}. Figure 4-1 is a schematic of the differential volume (DV) in a cylindrical pipe. Assuming that flow is in the axial direction alone, a mass balance for this control volume (CV) gives:

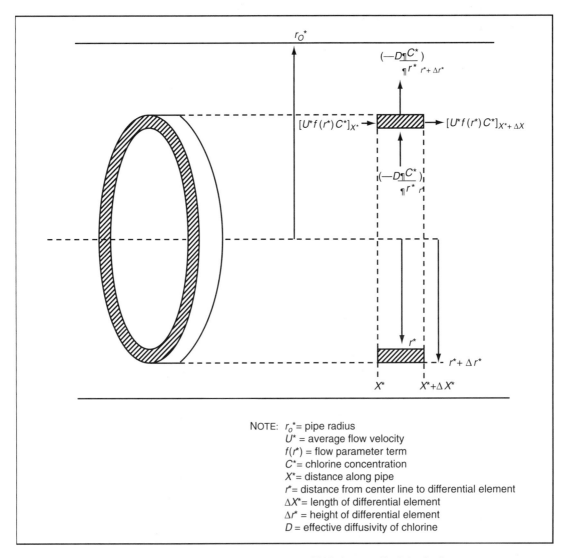

NOTE: r_o* = pipe radius
$U*$ = average flow velocity
$f(r*)$ = flow parameter term
$C*$ = chlorine concentration
$X*$ = distance along pipe
$r*$ = distance from center line to differential element
$\Delta X*$ = length of differential element
$\Delta r*$ = height of differential element
D = effective diffusivity of chlorine

Figure 4-1   Schematic of the differential volume (DV) in a cylindrical pipe

$$\partial(2\Pi r^*\Delta r^*\Delta X^* C^*)/\partial t^* = (U^*f(r^*)C^*2\Pi\Delta r^*)_{X^*} \qquad \text{(Eq 4-18)}$$
$$- (U^*f(r^*)C^*2\Pi r^*\Delta r^*)_{X^*+\Delta X^*} + (-D(\partial C^*/\partial X^*)2\Pi r^*\Delta r^*)_{X^*}$$
$$- (-D(\partial C^*/\partial X^*)2\Pi r^*\Delta r^*)_{X^*+\Delta X^*} - (-D(\partial C^*/\partial X^*)2\Pi r^*\Delta r^*)_{X^*+\Delta X^*}$$
$$- (-D(\partial C^*/\partial r^*)2\Pi r^*\Delta X^*)_{r^*} - (D(\partial C^*/\partial r^*)2\Pi r^*\Delta X^*)_{r^*+\Delta r^*}$$
$$- C^*2\Pi r^*\Delta r^*\Delta X^*$$

Where:

U^* = the average flow velocity throughout the distribution system, in cm/s

$f(r^*)$ = the flow parameter term depending on the flow regime. For laminar flow: $f(r^*) = 2[1 - (r^*/r_0^*)^2]$; for plug flow: $f(r^*) = 1$

r_0^* = the pipe radius, in cm

D = the effective diffusivity of chlorine species in water, in cm^2/s

k = the first-order decay rate constant in the bulk water (sec^{-1})

r^* = the radial distance from the center of the pipe, in cm

X^* = the axial distance from the inlet along the pipe, in cm

Δr^* and ΔX^* = incremental distances at r^* and X^*, respectively

C^* = the concentration of chlorine

The term on the left-hand side denotes a net accumulation of chlorine in the DV with time, the first two terms on the right-hand side account for increases in concentration due to the inflow of water into the DV, the next two terms account for increases due to diffusion of chlorine into the DV in the axial direction, the fifth and sixth terms account for increases due to diffusion of chlorine into the DV in the radial direction, and the last term on the right side accounts for the reduction of chlorine in the DV due to consumption by both chemical and microbiological reactions in the bulk water. Dividing the left side and right side by $2\Pi r^*\Delta r^*\Delta X^*$ and taking limits as Δr^* and ΔX^* go to zero, Eq 4-14 becomes

$$\partial C^*/\partial t^* \qquad \text{(Eq 4-19)}$$
$$= -\partial(U^*f(r^*)C^*)/\partial X^* + \partial(D\partial C^*/\partial X^*)/\partial X^* +$$
$$(1/r^*)(\partial(r^*D\partial C^*/\partial r^*)/r^*) - kC^*$$

where the variables are as previously defined.

Assuming quasi-steady conditions ($\partial C^*/\partial t^* = 0$) and rearranging Eq 4-19, we have

$$\partial(U^*f(r^*)C^*)/\partial X^* = \partial(D\partial C^*/\partial X^*)/\partial X^* \qquad \text{(Eq 4-20)}$$
$$+ (1/r^*)(\partial(r^*D\partial C^*/\partial r^*)/\partial r^*) - kC^*$$

The boundary conditions are as follows

$$X* = 0, \ C* = C_0* \text{ at } 0 \leq r* \leq r* \text{ entry conditions} \qquad \text{(Eq 4-21)}$$
$$r* = 0, \ \partial C*/\partial r* = 0 \text{ at } 0 \leq X* \leq L* \text{ symmetry condition}$$
$$r* = r_0*, \ D\partial C*/\partial r* = -V_d*C* \text{ at } 0 \leq X* \leq L* \text{ wall condition}$$

Where:

C_0* = the inlet chlorine concentration, in mg/L

$L*$ = the pipe length, in cm

V_d* = an empirical parameter proportional to the degree of absorbtion of the pipe surface, the "consumption rate" of chlorine at the pipe wall, in cm/s

The consumption process at the pipe wall can be compared to the transfer of mass or heat from the bulk liquid phase to the pipe surface. The following three wall conditions may be used, depending on the wall characteristics:

1. Perfect sink, $V_d* \to \infty$, i.e., $C* = 0$

2. No consumption at pipe wall, $V_d* = 0$, i.e., $\partial C*/\partial r* = 0$

3. Partial consumption, V_d* is some finite (nonzero) value

Assuming that the time period over which the experiments are done, $U*$ and D remain constant (quasi-steady state), and on nondimensionalizing Eq 4-20

$$\frac{f(r)\partial C}{\partial X} = \left(\frac{1}{Pe_a}\right)\left(\frac{\partial^2 C}{\partial X^2}\right) + \frac{A_0}{r}\frac{\partial}{\partial r}\left(r\frac{\partial c}{\partial r}\right) - A_1 C \qquad \text{(Eq 4-22)}$$

Where:

$$C = \frac{C*}{C_0*}$$

$$X = \frac{X*}{L*}$$

$$r = \frac{r*}{r_0^*}$$

$$Pe_a = \frac{L*U*}{D}$$

$$A_0 = \frac{L*D}{r*^2 U*} = \frac{\Pi L*D}{Q} \quad (Q = \text{flow rate throughout the system})$$

$$A_1 = \frac{kL*}{U*}$$

For distribution systems, the axial Peclet number Pe_a is typically large ($>10^6$), thus the axial diffusion term can be neglected with respect to the axial convective term. Turbulent flow conditions are often encountered, and then $f(r)$ is equal to 1, or the axial velocity can be assumed to be independent of the radial location for most parts of the cross section. Representing the axial velocity by a constant [$f(r)=1$] average value ($U*$), the governing equation for chlorine transport in pipes under these conditions is

$$\frac{\partial C}{\partial x} = \left(\frac{A_0}{r}\right)\left(\partial \frac{r \partial C}{\partial r}\right) - A_1 C \qquad \text{(Eq 4-23)}$$

with the boundary conditions

$$X = 0, \ C = 1, \text{ at } 0 \leq r \leq 1$$

$$r = 0, \ \frac{\partial C}{\partial r} = 0, \text{ at } 0 \leq X \leq 1$$

$$r = 1 \ \frac{\partial C}{\partial r} = -\left(\frac{V_d^* r_o^*}{D}\right)C \qquad \text{(Eq 4-24)}$$

$$= -A_2 C \text{ at } 0 \leq X \leq 1$$

Three dimensionless parameters A_0, A_1, and A_2 govern the chlorine decay in the distribution system. The parameter A_0 accounts for the radial diffusion and depends on the pipe length, on the effective diffusivity of chlorine, and on the flow rate throughout the system. The variable A_1 depends on the reactivity of chlorine with species such as viable cells or chemical compounds in the bulk liquid phase and on the residence time in the system. The variable A_2 is a wall consumption parameter depending on the wall consumption rate, on the pipe radius, and on the effective diffusivity of chlorine.

Since the axial velocity is represented by an average quantity, Eq 4-23 can be solved analytically. Using separation of variables, a solution to Eq 4-23 is as follows:

$$C(X, r) = 2\Sigma_{n=1}^{\infty} \frac{\lambda_n J_o(\lambda_n r)J_1(\lambda_n)}{(\lambda_n^2 + A_2^2)J_o^2(\lambda_n)} \times \exp\left[-(A_1 + \lambda_n^2 A_0)X\right] \qquad \text{(Eq 4-25)}$$

Where:

J_0 and J_1 = the Bessel functions of the first kind of order zero and 1, respectively

$\lambda_n s$ = the roots of $\lambda_n J_1(\lambda_n) = A_2 J_0(\lambda_n)$

Through a series of simplifying assumptions to Eq 4-25

$$C_{av} = \frac{\exp(-A_1 X)}{(1 + \varepsilon)} \qquad \text{(Eq 4-26)}$$

In Eq 4-22, an expression for ε for A_0 in the range of 10^{-4} to 100 is given by

$$\varepsilon = 2.4416\, A_0 A_2 - 0.1559\, A_0 A_2^2 \qquad 0.01 \leq A_2 < 10 \qquad (\text{Eq } 4\text{-}27)$$
$$\varepsilon = 10.105\, A_0 + 0.0014\, A_2 + 0.31\, A_0^2 A_z \qquad 0.01 \leq A_2 < 10$$

The average concentration at any location can be determined using Eq 4-26 and Eq 4-27 provided the three nondimensional parameters A_0, A_1, and A_2 are known.

Once the flow rates throughout the system are known (as can be computed provided there is an effective diffusivity of chlorine in water), D is also known. The value of D depends on ionic and molecular transport and on the flow regime. In actual practice, ionic transport has to be accounted for to determine the diffusivity (Lu 1991). If turbulent flow conditions prevail, the eddy diffusivity has to be used as it typically is greater than the molecular diffusivity ($D = 1.25 \times 10^{-5}$ cm^{2}s^{-1}).

Field data can be used to compute the three nondimensional parameters A_0, A_1, and A_2. Concentrations would need to be measured in at least three locations (in addition to the inlet), and Eq 4-26 and Eq 4-27 can then be solved to determine A_0, A_1, and A_2. If data are available at more than three locations, a least squares procedure can be used to determine the best fit values for A_0, A_1, and A_2.

DETERMINATION OF PARAMETERS FROM FIELD DATA

As described earlier, researchers have used a simple decay rate expression (as in Eq 4-17) to fit field data and report values of k. The value for k is a composite rate consumption that combines the effect of bulk liquid and wall consumption. As wall consumption and diffusion are not accounted for, different values of k are reported for different conditions. The model developed is also demonstrated by solving the equations for the conditions of a field sampling study conducted in the South Central Connecticut Regional Water Authority (SCCRWA). A schematic of the distribution system composed of 8-in. (200-mm) and 12-in. (300-mm) mains is shown in Figure 4-2.

The Brushy Plains/Cherry Hill service area operates under two basic scenarios—pumps on, when the tank is filling; and pumps off, when the tank is emptying. These scenarios hydraulically define links of pipe over which chlorine consumption can be measured. Chlorine concentrations at the inlet and outlet points of these pipe segments are shown in milligrams per litre in Table 4-2.

Five unidirectional flow segments were identified (shown in Table 4-2) wherein inlet and outlet chlorine concentrations were measured. The geometrical and flow parameters are listed in Table 4-3. As indicated in Table 4-3, the flow conditions in the pipe are turbulent, and hence the eddy diffusivity was used to estimate the diffusion coefficient, $D_{\text{eddy}} = 1.233 \times 10^{-2} U^* r_0^*$ at 78°F (25°C). Using these parameters, A_0 is computed for the different pipes and the

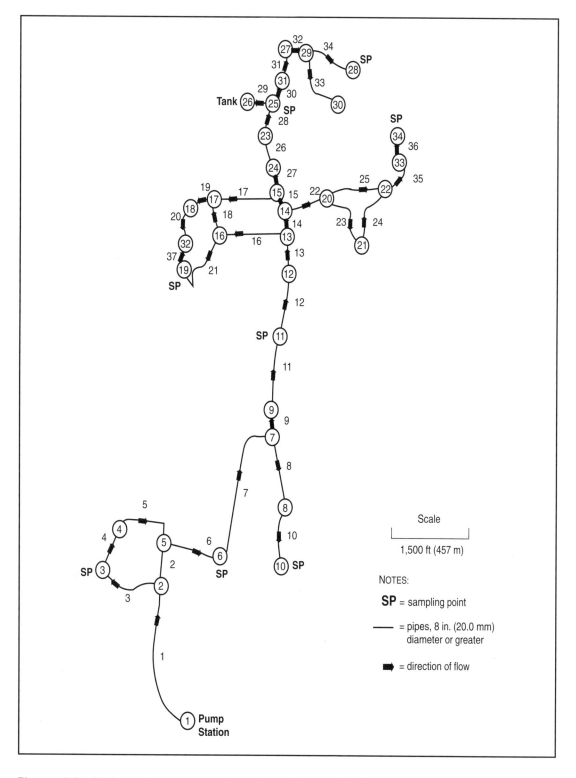

Figure 4-2   Link–node representation of the Cherry Hills/Brushy Plains network with "pumps-on" scenario

values listed in Table 4-3, k was estimated to be $6.4 \times 10^{-6} \text{s}^{-1}$ from bench kinetic tests performed with water samples taken at the inlet to the network. Assuming that the k value remains the same for the pipes, A_1 was computed for the different pipes and the values are listed in Table 4-3 (Biswas, Lu, and Clark 1993). This assumption is expected to be valid for the main network branch, however, it was not validated by performing a kinetic test with a water sample taken from a downstream sampling point. With A_1 and A_2 known, an iterative procedure using Eq 4-26 and Eq 4-27 was used to compute V_d^* and ε. The calculation is demonstrated for the segment with pipes 7, 9, and 11. The superscripts on the variables indicate the pipe number. At the outlet of the pipe, $X - 1$, and Eq 4-26 gives

$$C_{av}^7 = \frac{\exp(-A_1^7)}{(1 + \varepsilon^7)}$$

$$C_{av}^9 = \frac{\exp(-A_1^9)}{(1 + \varepsilon^9)} \qquad \text{(Eq 4-28)}$$

$$C_{av}^{11} = \frac{\exp(-A_1^{11})}{(1 + \varepsilon^{11})}$$

The measured concentration at the outlet to pipe 11 divided by the concentration at the inlet to pipe 7 is

$$C_{av}^7 C_{av}^9 C_{av}^{11} = \frac{0.98}{1.00}$$

$$= \frac{\exp(-A_1^7) \exp(-A_1^9) \exp(-A_1^{11})}{(1 + \varepsilon^7)(1 + \varepsilon^9)(1 + \varepsilon^{11})} \qquad \text{(Eq 4-29)}$$

The unknowns, ε^7, ε^9 and ε^{11} in Eq 4-29 are expressed as a function of A_0 and A_2 using Eq 4-27. The A_0s are known (Table 4-3) and A_2 can be expressed as a function of the unknown chlorine consumption rate at the pipe wall, V_d^*, and other known parameters. Similar expressions are developed for the other segments listed in Table 4-3. The value V_d^* is expected to be the same for pipes made of the same material in the main branch of the network. A spreadsheet program used wherein values of V_d^* could be entered and then A_2 and ε

Table 4-2   Chlorine concentrations at inlet and outlet of various segments

Pipes in Segment	Chlorine Concentration at Segment, in *mg/L*	
	Inlet	Outlet
1 and 3	1.08	1.00
7, 8, and 10	1.00	0.32
7, 9, and 11	1.00	0.98
12, 13, 16, and 21	0.98	0.16
12, 13, 14, 15, 26, 27, and 28	0.98	0.94

computed Eq 4-27. The outlet concentrations were then estimated using expressions similar to Eq 4-28, and then compared to the measured concentration ratios (Table 4-3). A value of $V_d^* = 1 \times 10^{-7} ms^{-1}$ for the main branch pipes led to good agreement with the measurements. However, different values of V_d^* were obtained for the pipes on dead ends or off the main network branch (pipes 3, 8, 10, and 21). Pipes 3, 8, 10, and 21 have high ε values indicating that wall consumption is significant. These are off-branch pipes (Figure 4-1) and qualitative observations indicate that significant biofilm growth occurs on such pipes.

QUALNET

Islam (1995) developed a model called QUALNET, which predicts the spatial and temporal distribution of chlorine residuals in pipe networks under slowly

Table 4-3   Model parameters for different pipes in the network

Pipe	Length, m	Radius, m	Flow Velocity, m/s	Diffusion Coefficient, m^2/s	Reynolds Number	V_d^*, m/s	A_0	A_1	A_2	ε
1	731.5	0.152	0.546	1.03e–3	1.75e+5	1.00e–7	5.92e+1	8.58e–3	1.49e–5	2.2e–3
3	396.2	0.102	0.195	2.46e–4	4.19e+4	1.15e–6	4.79e+1	1.30e–2	4.78e–4	5.6e–2
7	822.9	0.152	0.512	9.62e–4	1.64e+5	1.00e–7	6.66e+1	1.03e–2	1.58e–5	2.6e–3
8	365.8	0.152	0.014	2.58e–5	4.40e+3	1.00e–7	2.96e+1	1.71e–1	5.91e–4	4.3e–2
9	121.9	0.152	0.494	9.28e–4	1.58e+5	1.00e–7	9.86e+0	1.58e–3	1.64e–5	4.0e–4
10	304.8	0.102	0.014	1.80e–5	3.08e+3	2.3e–6	3.68e+1	1.36e–1	1.30e–2	1.17
11	213.4	0.152	0.485	9.11e–4	1.55e+5	1.00e–7	1.73e+1	2.82e–3	1.67e–5	7.1e–4
12	579.1	0.152	0.457	8.59e–4	1.47e+5	1.00e–7	4.69e+1	8.11e–3	1.77e–5	2.0e–3
13	182.9	0.152	0.445	8.36e–4	1.43e+5	1.00e–7	1.48e+1	2.63e–3	1.82e–5	6.6e–4
14	121.9	0.152	0.372	6.99e–4	1.19e+5	1.00e–7	9.86e+0	2.10e–3	2.18e–5	5.3e–4
15	91.4	0.152	0.329	6.19e–4	1.06e+5	1.00e–7	7.40e+0	1.78e–3	2.46e–5	4.5e–4
16	457.2	0.102	0.168	2.11e–4	3.60e+4	1.00e–7	5.52e+1	1.75e–3	4.84e–5	6.5e–3
21	426.7	0.102	0.049	6.14e–5	1.05e+4	2.2e–5	5.15e+1	5.60e–2	3.66e–2	4.589
26	182.9	0.152	0.329	6.19e–4	1.06e+5	1.00e–7	1.48e+1	3.56e–3	2.46e–5	8.9e–4
27	76.2	0.152	0.338	6.36e–4	1.09e+5	1.00e–7	6.17e+0	1.44e–3	2.40e–5	3.6e–4
28	91.4	0.152	0.323	6.07e–4	1.04e+5	1.00e–7	7.40e+0	1.81e–3	2.51e–5	4.5e–4

Source: From Biswas, Lu, and Clark (1993).

varying unsteady flow conditions. Unlike other available models, which use steady state or extended-period simulation (EPS) of steady flow conditions, QUALNET used a lumped-system approach to compute unsteady flow conditions and includes dispersion and decay of chlorine during travel in a pipe. The pipe network is first analyzed to determine the initial steady state conditions. The slowly varying conditions are then computed by numerically integrating the governing equations by an implicit-finite-difference, a scheme subject to the appropriate boundary conditions. The one-dimensional dispersion equation is used to calculate the concentration of chlorine over time during travel in a pipe, assuming a first-order decay rate.

Numerical techniques are used to solve the dispersion, diffusion, and decay equation. Complete mixing is assumed at the pipe junctions. The model has been verified by comparing the results with those of EPANET for two typical networks. The results are in good agreement at the beginning of the simulation model for unsteady flow; however, chlorine concentrations at different nodes vary when the flow becomes unsteady and when reverse flows occur. The model may be used to analyze the propagation and decay of any other substance for which a first-order reaction rate is valid. The dynamic equation describing the slowly time-varying flow in a pipe may be written as follows (Chaudhry 1987):

$$\frac{L}{gA}\frac{dQ}{dt} = H_1 - H_2 - kQ|Q^m \qquad \text{(Eq 4-30)}$$

Where:

L = pipe length

g = acceleration due to gravity

A = pipe cross-sectional area, in ft^2 (m^2)

Q = rate of discharge, in ft/sec (m/sec)

t = time

H_1 = piezometric head on the upstream end of pipe, in ft (m)

H_2 = piezometric head on the downstream end of pipe, in ft (m)

k (L/sec) and m = constants in a head loss formula of an exponential type (i.e., Darcy–Weisbach or Hazen–Williams).

The absolute value allows for a proper sign for the head loss term in case the flow reverses during the simulation.

The continuity equation describing the mass conservation at a junction of pipes may be written as

$$\Sigma Q_j + \Sigma Q_0 = 0 \qquad \text{(Eq 4-31)}$$

in which the inflows are considered positive and the outflows are considered negative.

A computational technique developed by Holloway (1985) is used to solve the simulation of slowly varying flows in a pipe. For a first-order decay rate, the one-dimensional dispersion of a constituent in a pipe is modeled by the following equation:

$$\frac{\partial C}{\partial t} + V\frac{\partial C}{\partial x} = D\frac{\partial^2 C}{\partial x^2} - k_1 C \qquad \text{(Eq 4-32)}$$

Where:

C = constituent concentration

V = advective velocity, in ft/sec (m/sec)

D = dispersion coefficient, in ft^2/sec (m^2/sec)

x = spatial coordinate along the pipe axis, in ft (m)

k_1 = first-order reaction rate coefficient, in sec^{-1}

In order to avoid problems with numerical dispersion in solving Eq 4-32, it is split into the following two separate equations:

$$\frac{\partial C}{\partial t} + V\frac{\partial C}{\partial z} = 0 \qquad \text{(Eq 4-33)}$$

$$\frac{\partial C}{\partial t} - D\frac{\partial^2 C}{\partial x^2} - k_1 C = 0 \qquad \text{(Eq 4-34)}$$

Equation 4-29 can be discretized first to convert the mass of constituents a certain distance downstream for each time step so that there is no numerical diffusion. Then the advected concentrations are diffused longitudinally by solving Eq 4-34 at each spatial grid point, including their decay. The final concentration field is thus obtained from the entire time increment. The diffusion coefficient was estimated by the following equation:

$$D = 10.1\tau\sqrt{\frac{\tau_0}{\rho}}\left(\frac{\text{cm}^2}{\text{sec}}\right) \qquad \text{(Eq 4-35)}$$

Where:

τ = pipe radius

τ_0 = shear stress at the wall

ρ = density of the fluid

The Warming–Kutler–Lomax scheme was used for the solution. QUALNET was calibrated on sample networks against EPANET. The results compared favorably.

Event-Driven Method

The water quality simulation process used in the previous models is based on a one-dimensional transport model, in conjunction with the assumption that complete mixing of material occurs at the junction of pipes. These models consist of moving the substance concentrations forward in time at the mean flow velocity while undergoing a concentration change based on kinetic assumptions. The simulation proceeds by considering all the changes to the state of the system as the changes occur in chronological order. Based on this approach, the advective movement of substance defines the dynamic simulation model. Most water quality simulation models are interval oriented, which in some cases, can lead to solutions that are either prohibitively expensive or contain excessive errors.

Boulos et. al (1995) proposed a technique mentioned earlier called the event driven method (EDM). This is extremely simple in concept and is based on a next-event scheduling approach. In this method, the simulation clock time is advanced to the time of the next event to take place in the system. The simulation scheduled is executed by carrying out all the changes to a system associated with an event, as events occur in chronological order. Since the only factors affecting the concentration at any node are the concentrations and flows at the pipes immediately upstream of the given node, the only information that must be available during the simulation are the different segment concentrations. The technique makes the water quality simulation process very efficient.

The advective transport process is dictated by the distribution system demand. The model follows a front tracking approach and explicitly determines the optimal pipe segmentation scheme with the smallest number of segments necessary to carry out the simulation process. To each pipe, pointers (concentration fronts), whose function is to delineate volumes of water with different concentrations, are dynamically assigned. Particles representing substance injections are processed in chronological order as they encounter the nodes. All concentration fronts are advanced within their respective pipes based on their velocities. As the injected constituent moves through the system, the position of the concentration fronts defines the spatial location behind which constituent concentrations exist at any given time. The concentration at each affected node is then given in the form of a time–concentration histogram.

The primary advantage of this model is that it allows for dynamic water quality modeling that is less sensitive to the structure of the network and to the length of the simulation process. In addition, numerical dispersion of the concentration front profile resolution is nearly eliminated. The method can be readily applied to all types of network configurations and dynamic hydraulic conditions and has been shown to exhibit excellent convergence characteristics.

References

Biswas, P., C. Lu, and R.M. Clark. 1993. A Model for Chlorine Concentration Decay in Drinking Water Distribution Pipes. *Water Research*, 27(12):1,715–1,724.

Boulos, P.F., T. Altman, P.A. Jarrige, and F. Collevati. 1995. Discrete Simulation Approach for Network-Water-Quality Models. *Jour. Water Resources Planning and Management*, 121(1):49–60.

Chaudhry, M.H. 1987. *Applied Hydraulic Transients*, 2nd ed. New York: Van Nostrand Reinhold.

Lu, C. 1991. Theoretical Study at Particle, Chemical, and Microbial Transport In Drinking Water Distribution Systems. PhD thesis. Cincinnati, Ohio: University of Cincinnati.

Islam, M.R. 1995. Modeling of Chlorine Concentration in Unsteady Flows in Pipe Networks. PhD thesis. Pullman, Wash.: Washington State University.

Holloway, M.B. 1985. Dynamic Pipe Network Computer Model. PhD thesis. Pullman, Wash: Washington State University.

Rossman, L.A. 1994. EPANET User's Manual. Cincinnati, Ohio: USEPA.

Rossman, L.A., R.M. Clark, and W.M. Grayman. 1994. Modeling Chlorine Residuals in Drinking Water Distribution Systems. *Jour. Environ. Eng.*, 120(4):803–820.

Initial Modeling Studies

One of the first comprehensive projects designed to investigate the feasibility of water quality modeling development and application was a cooperative agreement initiated between the North Penn Water Authority (NPWA), Lansdale, Pa., and the US Environmental Protection Agency (USEPA) to study the mixing of water from multiple sources. This project investigated the feasibility for development and application of a steady state water quality model. As the study progressed, it became obvious that the dynamic nature of both demand patterns and water quality variations required the development of a dynamic water quality model. In addition, techniques for semicontinuous monitoring of volatile organic contaminants (VOCs) were explored (Clark et al. 1988).

This chapter discusses the historical development of water quality models as applied to drinking water distribution systems. It describes the application of water quality models to two case study utilities—the NPWA and the South Central Connecticut Regional Water Authority (SCCRWA).

North Penn Case Study

At the time of the study, the NPWA served 14,500 customers in 10 municipalities. Average demand was 5 mgd (19 ML/d) (Clark et al. 1988). Water sources included a 1-mgd (3.7-ML/d) treated surface water source purchased from the Keystone Water Company and 4 mgd (15 ML/d) from 40 wells operated by NPWA. Figure 5-1 is a schematic representation of the 225 mi (362 km) of pipe in the NPWA distribution system. The figure shows the location of wells, the Keystone tie-in, and the three pressure zones—the Souderton zone, Lansdale low zone, and Hillcrest zone.

Surface water entered the NPWA system at the Keystone tie-in. The rate of flow into the system was determined by the elevation of the tank in the Keystone system, and by a throttling valve at the tie-in. Flow was monitored continuously, and was relatively constant. Water flowed into the Lansdale low-pressure zone, entered the Lawn Avenue tank, and was pumped into the Souderton zone. Additional water from the Hillcrest pressure zone entered the Lansdale system at the Office Hillcrest transfer point. This water was solely derived from wells in the Hillcrest zone. Except for unusual and extreme circumstances, such as fire or main breaks, water did not flow from the Souderton zone into the Lansdale low zone, nor from Lansdale low into Hillcrest.

There were distinct chemical characteristics associated with the Keystone water as compared with the well waters. Keystone water contained total trihalomethanes (TTHM) at significantly higher levels than the well water. Certain wells showed the occurrence of trichloroethylene (TCE) and/or cis-1,2-dichloroethylene (cis-1,2-DCE). Inorganic chemicals also varied from well to well, and between the wells and Keystone.

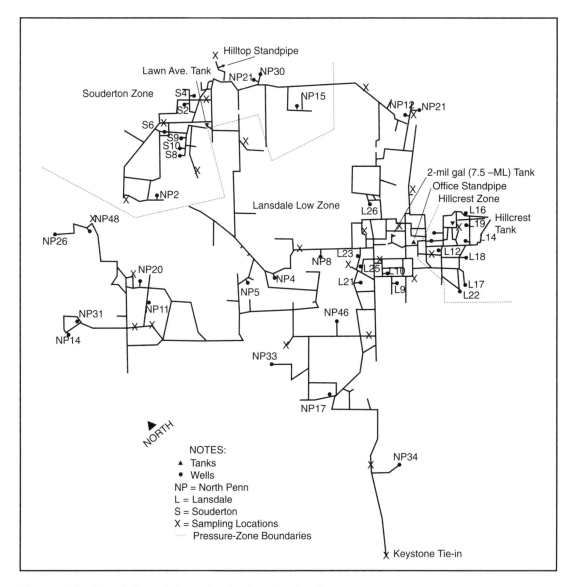

Figure 5-1 North Penn Water Authority distribution system

NETWORK MODELING

The NPWA distribution system was modeled in a network representation consisting of 528 links and 456 nodes. Water demands for modeling represented conditions from May–July 1984. The network hydraulic model used was the WADISO model, which contained provisions for both steady state and quasi-dynamic hydraulic modeling or EPS (Gessler and Walski 1985).

The model was employed to study the overall sensitivity of the system to well pumpage, demand, and other factors resulting in the development of a number of typical flow scenarios.

The modeling effort revealed that significant portions of the system were subject to flow reversals. This type of analysis was used to select sampling points for an in-depth sampling effort described later.

DYNAMIC VARIATIONS IN WATER QUALITY DATA

To investigate the nature of water quality variability under dynamic conditions within the system, a sampling program was conducted at six sites. Site selection was based on spatial variations determined from historical data and modeling results. Figure 5-2 shows the various sampling points used in the field study. Figure 5-3 depicts the results of the intensive sampling program using total trihalomethanes (TTHMs) as a tracer. Laboratory and field evaluations demonstrated that the THMs in the Keystone water had reached their formation potential and were relatively stable. Any trihalomethanes (THMs) formed from well sources were relatively minor. Figure 5-3 also shows the variation in hardness of these same points (hardness was primarily associated with flow from the wells). At the Mainland sampling point, a flushing back and forth of water between the surface source and the well sources could be seen. The peaks of the TTHMs at Mainland were approximately 12 hours out of phase with the peaks from the wells. This indicates that water flow at this point was affected by the surface and groundwater sources. The results pointed out the problems in attempting to predict a dynamic situation using a steady state approach and the dynamic nature of water movement in the distribution system.

DYNAMIC WATER QUALITY ALGORITHM DEVELOPMENT

A dynamic water quality model was developed using a numerical routing solution to trace water quality through the network. Demands and inflows (both values and concentrations) were assumed constant over a user-defined period and a quasi-dynamic externally generated hydraulic solution was used for each period. Thus, the flow and velocity for each link was known from the hydraulic solution for each time period. Each time period was evenly divided into an integer number of computational time steps. Each link was then divided into sublinks by a series of evenly spaced subnodes (though the distance between subnodes varied from link to link or for a link at different time periods), such that the travel time from a subnode (or node) to the adjacent subnode (or node) was approximately equal to Δt.

The information required by the dynamic model was classified into the following three categories: general information, initial conditions, and information required for each time period. This required information is summarized in Table 5-1.

The solution algorithm used in the dynamic water quality model was operated sequentially by time period. During a time period, all external forces affecting the water quality were assumed to remain constant (i.e., demand, well pumpage, tank head, etc.).

An example of the movement of a contaminant is illustrated in Figure 5-4, in which darker shading indicates higher concentrations of contaminants, while lighter or white corresponds to low or zero concentration. As illustrated, during each time step, the contents of a subnode moves to the adjacent subnode

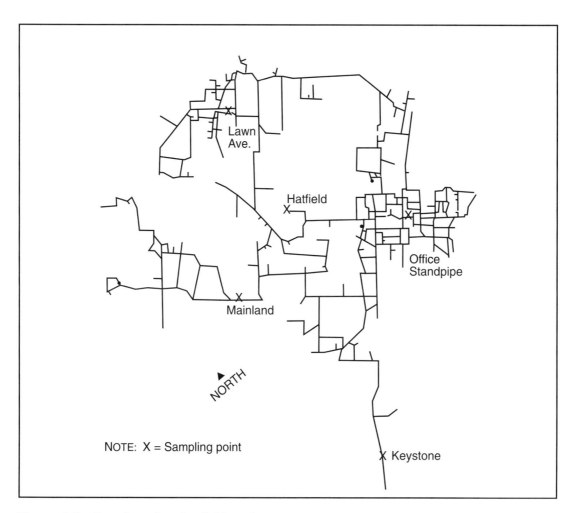

Figure 5-2 Sample points for field study

traveling in the direction of flow. In this example, at the end of a time period (i.e., after time step T3), the direction of flow changes and the number of sub-nodes change due to an increase in velocity. Accordingly, the high concentration packets change direction and move from right to left.

Table 5-1   Information required by the dynamic water quality model

General Information
- Δt time step

General Network Information
- node numbers associated with the end of each link
- link lengths
- pipe diameters
- node number associated with each source
- node number associated with each tank
- tank geometry

Initial Conditions
- concentration at each node at the start of simulation
- volume in tank at start of simulation

Information Required for Each Period
- direction and flow in each link
- velocity in each link (optionally may be calculated based on pipe diameter)
- concentration in source flow

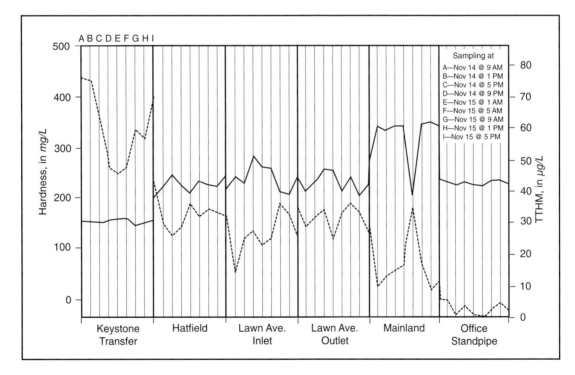

Figure 5-3   Results from six sampling stations in North Penn Water Authority

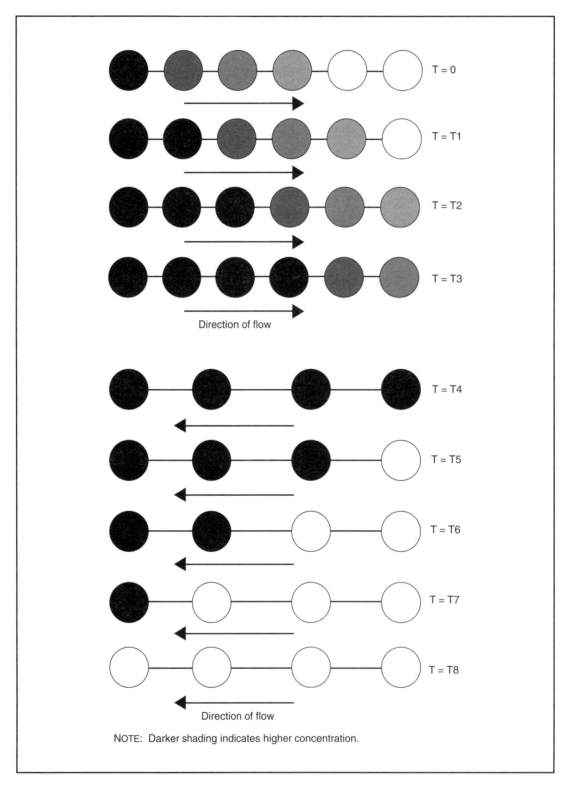

Direction of flow

Direction of flow

NOTE: Darker shading indicates higher concentration.

Figure 5-4   Example movement of contaminant within link

MODEL APPLICATION

The dynamic water quality algorithm was implemented as a microcomputer-based dynamic water quality model written in FORTRAN. Versions for both Apple Macintosh® and DOS-based computers were developed. The program, called the Dynamic Water Quality Model (DWQM), was applied in a full-scale demonstration on the NPWA distribution system.

A 34-hour period was simulated corresponding to conditions present during the pilot-level sampling program conducted on Nov. 14–15, 1985. Parameters of the model were adjusted so that predicted tank levels and flows at selected sites represented those measured during the sampling period. A comparison of measured and modeled hydraulic conditions at three locations are shown in Figure 5-5. The resulting predicted time history of chloroform is plotted in Figure 5-6 along with field sampling results for three sampling stations.

Modeling results for TTHM are plotted in Figure 5-7 for three sampling stations, as are the sampling results. The results of the predicted hardness as compared to sampling measurements are presented in Figure 5-8.

The results of applying the dynamic water quality model to the North Penn system as reflected in Figures 5-6 and 5-7 indicate both close agreement between predicted and observed results and some anomalous behavior. For chloroform, TTHM, and hardness, the general levels of concentration compare very favorably to the observed values at the three selected sampling stations (Figures 5-6 and 5-7). In each case, there were some differences in the timing of peak or minimum values. When the spatial variation of predicted TTHM concentrations were compared to the historical average TTHM level, the same general patterns were apparent. Additionally, the predicted patterns bracketed the pattern corresponding to the long-term historical average; a result that would be expected because the two selected times correspond to the extreme spatial patterns during the sampling period.

South Central Connecticut
Regional Water Authority Case Study

The North Penn case study, discussed previously, provided fertile ground for developing a dynamic water quality propagation model. In order to extend that application, USEPA initiated a cooperative agreement with the University of Michigan. Under the agreement, the university began a program with the South Central Connecticut Regional Water Authority (SCCRWA) to test the previously developed modeling concepts, including field studies to verify and calibrate the model (Clark and Goodrich 1993).

At the time of the study, the SCCRWA supplied water to approximately 95,000 customers (380,000 individuals) in 12 municipalities in the New Haven, Conn., area. The SCCRWA service area was divided into 16 separate pressure/distribution zones (see Figure 5-9). Average production was 50 mgd (189 ML/d) with a safe yield of approximately 74 mgd (280 ML/d). Surface water sources included Lake Gaillard, Lake Saltonstall, Lake Whitney, and the West River system. There were five well fields serving as sources (North Cheshire, South Cheshire, Mt. Carmel, North Sleeping Giant, and South Sleeping Giant). Approximately 80 percent of the water in use in the system

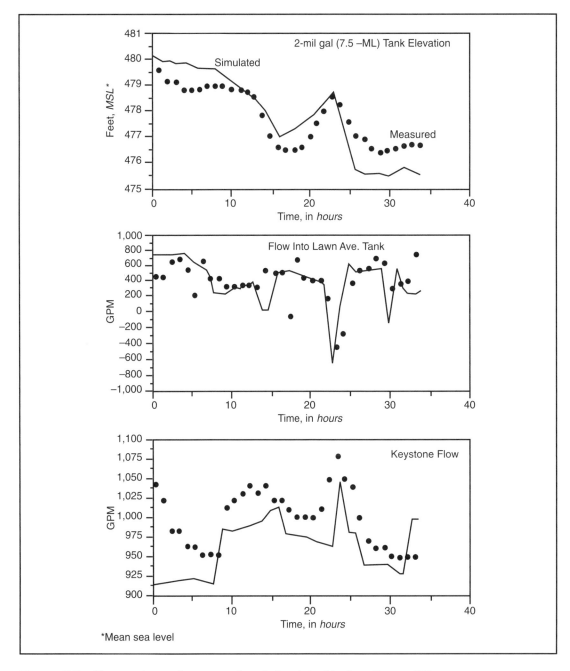

Figure 5-5   Comparison of measured and simulated hydraulic conditions

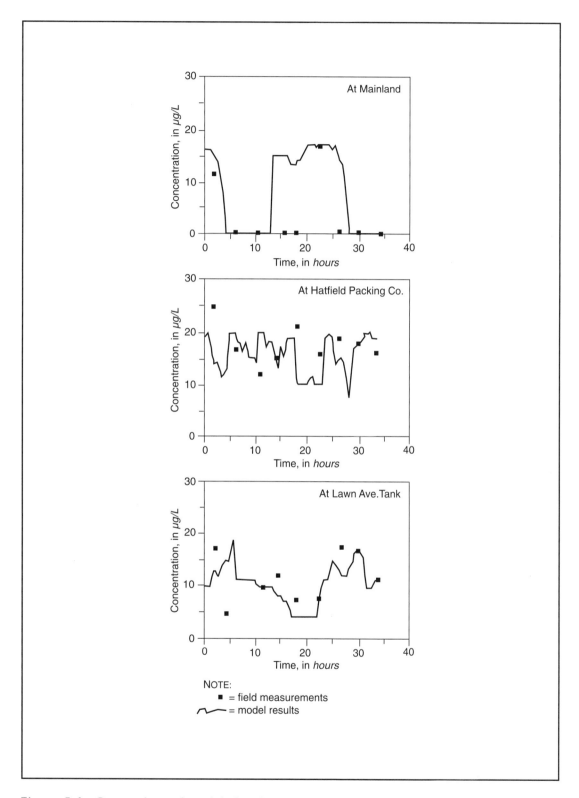

Figure 5-6   Comparison of modeled and measured chloroform at selected stations

came from surface sources, with the remaining 20 percent from wells. All water was treated using chlorination, filtration, and a phosphate corrosion inhibitor. The system included 22 pumping stations, 23 storage tanks, and approximately 1,300 mi (2,080 km) of water mains.

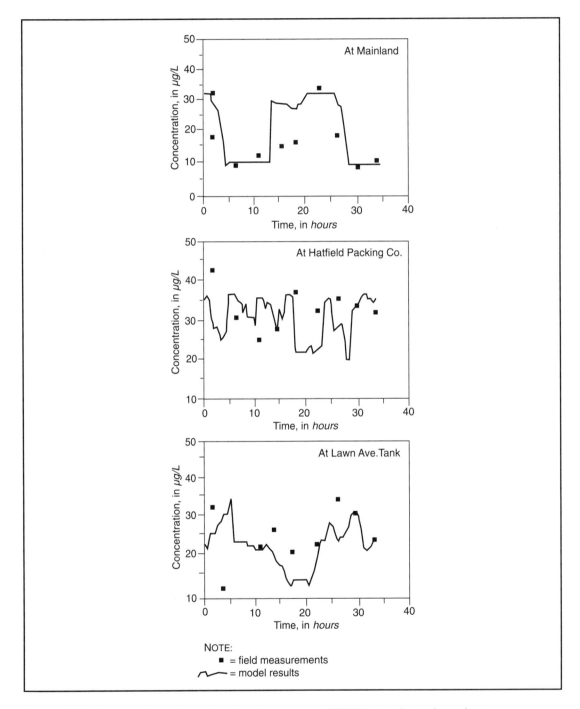

Figure 5-7 Comparison of modeled and measured TTHM at selected stations

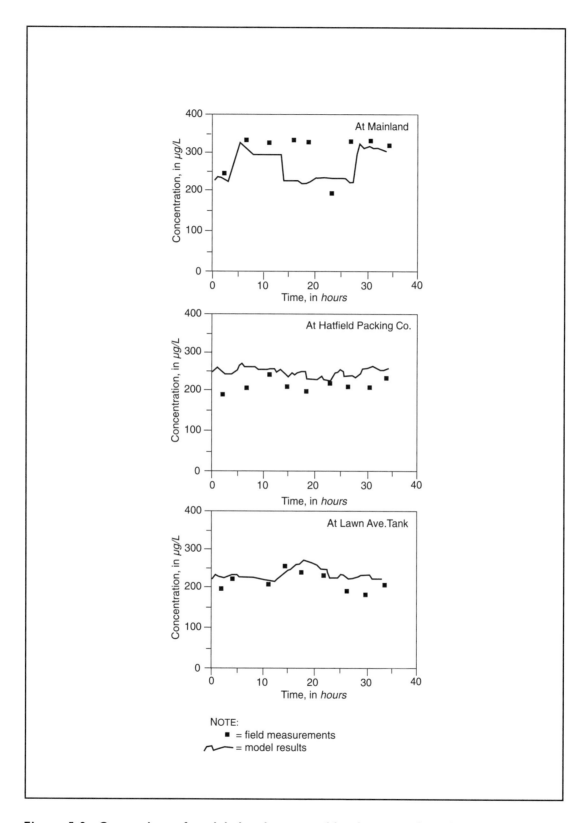

Figure 5-8 Comparison of modeled and measured hardness at selected stations

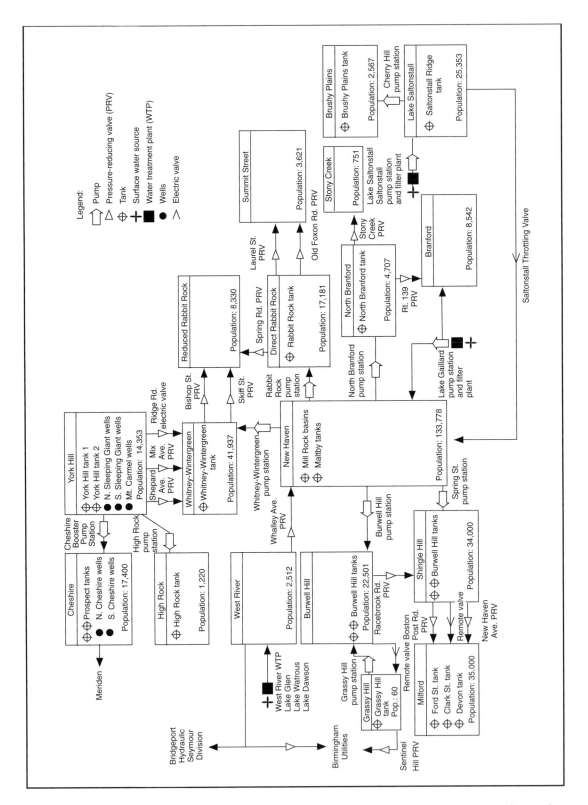

Figure 5-9 South Central Connecticut Regional Water Authority service area schematic

SYSTEM MODELING

The preliminary efforts in developing and validating a model for the SCCRWA were concentrated on the Cheshire service area. This area is relatively isolated and provided a prototype for modeling the remainder of the system. In order to validate the model, an extensive study was planned in which the fluoride feed at the North Well field was turned off and the propagation of the fluoride feed water tracked through the system.

HYDRAULIC MODELING

Prior to the application of the water quality modeling effort and related field study, extensive hydraulic analyses were conducted on the system. For the preliminary modeling, the full SCCRWA system network was represented by approximately 520 nodes and 700 links. This representation was modified based on the most current maps supplied by the authority. For other service areas, distribution system maps were used to develop a network skeletonization at a level of detail consistent throughout the system. The network was drawn as a mylar overlay to the US Geological Survey (USGS) 1:24,000-scale topographic maps. In most cases, the system was represented by skeletonization (i.e., selectively choosing pipes based on their size and perceived impact as transmission mains). In a few cases, surrogate pipes were used to represent the effects of several pipes. For example, a single 14-in. (355-mm) pipe may be used to represent the effects of a pair of parallel 8-in. (200-mm) and 12-in. (305-mm) pipes. In these model applications, the hydraulic and water quality models were applied to the Cheshire system simulating the proposed fluoride study.

The first step in the modeling process after database development and hydraulic model implementation was to apply the steady state and dynamic water quality models developed as a result of the North Penn study to the Cheshire system. This was designed to simulate the propagation of fluoride feed water and thus select sampling locations when the fluoride feed was actually turned off.

DESIGN OF THE FIELD PROGRAM

Based on the simulation results and the objectives of the proposed field study, a preliminary field testing program was designed. Considerable refinement was required based on actual availability of manpower, variations in the operation of the system, etc. However, the plan served as a starting point for final design of the sampling plan.

Fluoride was picked as the tracer for several reasons. It is regularly added to the water at a concentration of approximately 1 mg/L as required by the state Department of Health Services, it does not dissipate from the water, it is easily tested for, and turning off the fluoride feed could be done with no health or aesthetic effects to the water. A fluoride concentration between 0.8 and

1.2 mg/L in public drinking water is required by the state Department of Health Services' Public Health Code. The natural background concentration of fluoride in the Cheshire wells was approximately 0.10 mg/L. By tracing the changes in the fluoride concentration in the distribution system, accurate travel times could be determined by relating the time the fluoride feed was shut down at the well fields to the time it dissipated at the sampling points (Skov, Hess, and Smith 1991).

PREPARATION

Based on their strategic location in the distribution system, the suggested sites for continuous analyzers were at three elementary schools and the Cheshire town hall (Figure 5-10). Permission to use the sites was obtained via phone contact and official letters to the appropriate Town and Board of Education personnel. Each site was then visited to determine if there was a viable location in each building to set up a continuous analyzer and chart recorder. Location requirements included an outlet for power, a sample tap at or near the location where the water service entered the building, a drain to accept the analyzed water, and an area large enough to set up equipment.

RESULTS FROM THE FIELD STUDY

Results from the study were very encouraging. The model predictions for the propagation of the nonfluoridated water through the system were very close. Figures 5-11 and 5-12 are comparisons between the model predictions and the field sampling results for nodes 37, 88, 182, 323, 537, and 570 (tank). The general trends predicted were borne out by the field study. As can be seen from the bottom plot in Figure 5-12, the concentration in the tank discharge increased or decreased until it approached the calculated concentration level in the tank. This effect was probably due to a lack of complete mixing in the tank.

The behavior of the tanks is of particular interest. During the early part of the sampling period, tank level variations were held to a minimum (<3 ft [<0.9 m]). After two days, little change in fluoride concentrations was found in the tank and, as a result, the water level was then allowed to vary approximately 8 ft (2.4 m). The wider range in tank level variation had the effect of turning the water over relatively rapidly. Even with the rapid turn over it took nearly 10 days to fully replace old water with new water in the tanks. It was clear from this analysis that tanks could have a detrimental effect on water quality, particularly as water ages in the tank.

From Aug. 13–15, 1991, another sampling program at the Cherry Hill/Brushy Plains service area began to validate the previously discussed simulation results. The purpose of this sampling program was to gather information to characterize the variation of water quality in the service area and to study the impact of tank operation on water quality.

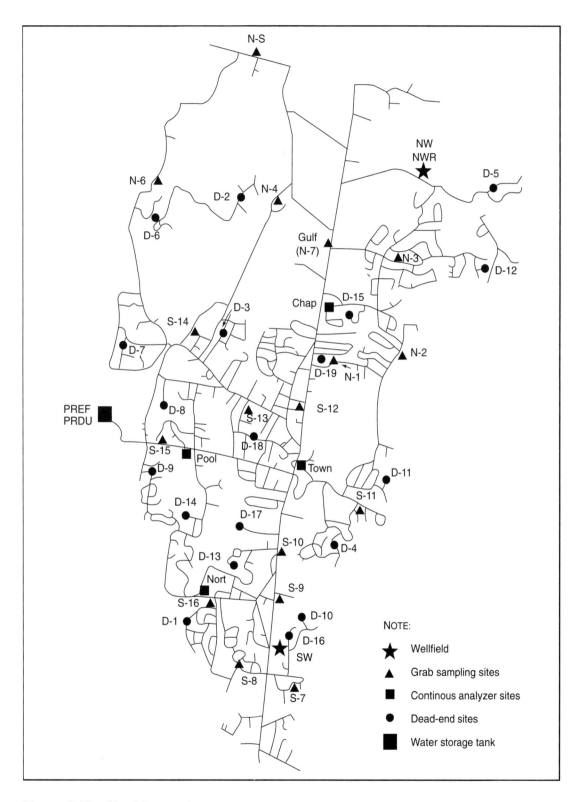

Figure 5-10   Cheshire service area map

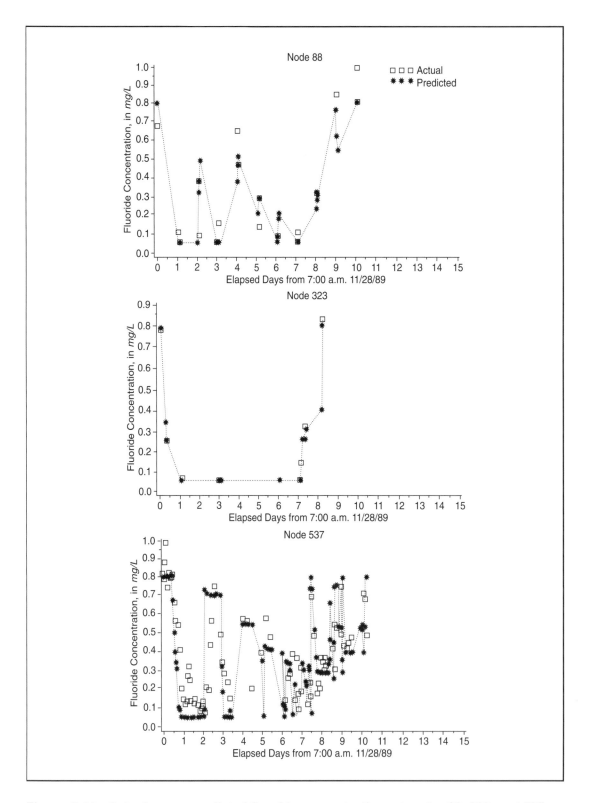

Figure 5-11   Actual versus predicted fluoride concentrations at nodes 88, 323, and 537

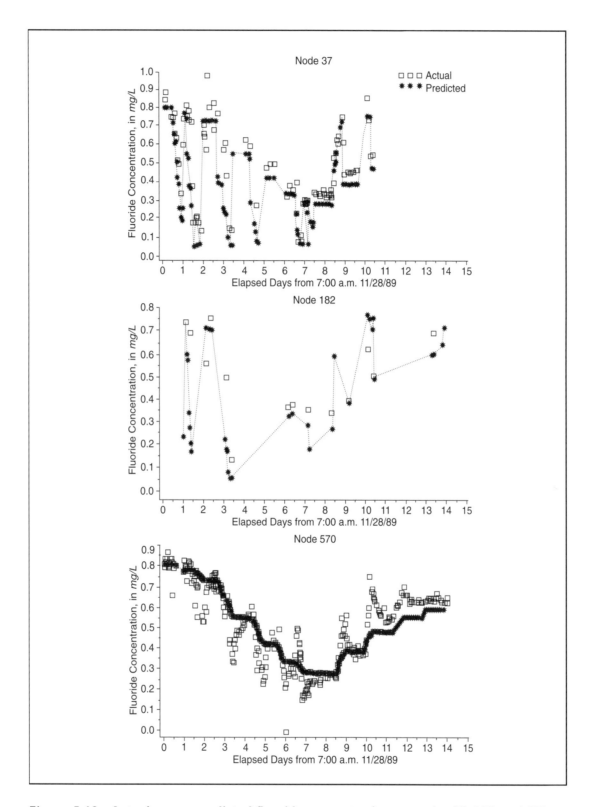

Figure 5-12 Actual versus predicted fluoride concentrations at nodes 37, 182, and 570

VERIFICATION STUDY

The Cherry Hill/Brushy Plains service area covered approximately 2 mi^2 (5 km^2) in the town of Branford in the eastern portion of the SCCRWA area (Clark et al. 1993). The service area was almost entirely residential, containing both single-family homes and apartment and condominium units. Average water use during the sampling period was 0.46 mgd (1,700 m^3/d). The water distribution system consisted of 8-in. (200-mm) and 12-in. (305-mm) mains as shown in the schematic in Figure 5-13. The terrain in the Cherry Hill/Brushy Plains service area was generally moderately sloping with elevations varying from approximately 50 ft (15.2 m) mean sea level (msl) to 230 ft (70.1 m) msl.

Cherry Hill/Brushy Plains received its water from the Saltonstall system. Water was pumped from the Saltonstall system into Brushy Plains by the Cherry Hill pump station. Within the service area, storage was provided by the Brushy Plains tank. The pump station contained two 4-in. (100-mm) centrifugal pumps with a total capacity of 1.4 mgd (5,300 m^3/d). The operation of the pumps was controlled by water elevation in the tank. Built in 1957, the tank had a capacity of 1.0 mgd (3,800 m^3/d). It had a diameter of 50 ft (15.2 m), a bottom elevation of 193 ft (58.8 m) msl, and a height (to the overflow) of 263 ft (80.2 m) msl. During normal operation, the pumps were set to go on when the water level in the tank dropped to 56 ft (15.2 m) and to turn off when the water level reaches 65 ft (19.8 m).

PRESAMPLING PROCEDURES

Prior to the sampling period, the WADISO hydraulic model (discussed on page 75) and the DWQM were applied to establish flow patterns within the service area. Additionally, during the periods of May 21–22, July 1–3, July 8–10, and July 30–Aug. 1, 1991, chlorine residuals were monitored (using a portable free chlorine analyzer and chart recorder) at the tank and operational patterns (pump records and tank water-level variations) were studied.

Based on these model runs and sampling data, a sampling strategy was adopted. This strategy involved turning the fluoride off at the Saltonstall treatment facility and sampling for both fluoride and chlorine in the Cherry Hill/Brushy Plains service area. Defluorided water was used as a conservative tracer for the movement of flow through the system and to calibrate the DWQM. The DWQM and a chlorine decay model based on hydrodynamic principles was used to model the dynamics of this substance. Seven sampling sites in the distribution system, in addition to sampling sites at the pump station and tank, were identified as shown in Figure 5-13.

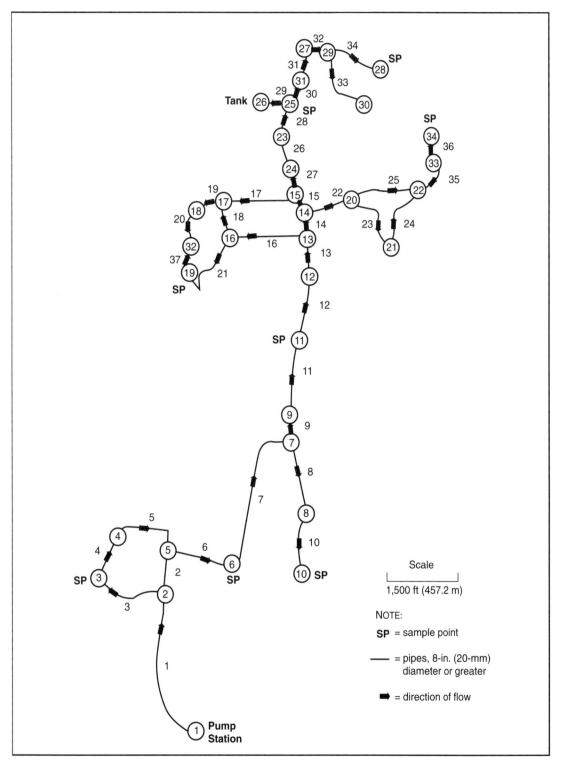

Figure 5-13   Link–node representation of the Cherry Hill/Brushy Plains network with "pumps-on" scenario

ANALYSIS OF SAMPLING RESULTS

The WADISO hydraulic model and DWQM water quality model were used to simulate the Cherry Hill/Brushy Plains service area for a 53-hour period from 9:00 a.m. on August 13 to 3:00 p.m. on Aug. 15, 1991. This skeletonization of the distribution system, as shown in Figure 5-13, included all 12-in. (305-mm) mains, major 8-in. (200-mm) mains and loops, and pipes that connected to the sampling sites. Pipe lengths were scaled from maps, actual pipe diameters used, and, in the absence of any other information, a Hazen–Williams roughness coefficient of 100 was assumed for all pipes.

Figures 5-14 and 5-15 show the results of the fluoride sampling study and the modeling efforts at each of the sampling nodes. From these results it is clear that the modeling effort matches the sampling efforts quite well with the exception of the dead ends. The shaded areas at the top of the individual plots in Figures 5-14 and 5-15 show the on-and-off cycle for the pumps. Clearly, the pump cycles heavily influence water quality at several sampling points. For example, at node 11, during the pumps' on cycle, the fluoridated water is pumped into the system. When the system is being fed from the tank (pumps off), the system is receiving water that had reached an equilibrium concentration of fluoride prior to the stoppage of the fluoride feeders.

CHLORINE RESIDUAL MODELING

Table 5-2 shows the length of pipe in feet, average velocity in feet per second, residence time in days, and the average chlorine residual at the beginning and ending nodes for various links in the system with the pumps on.

Using the upstream and downstream chlorine concentration and the residence times in the link, the chlorine decay coefficient was calculated for each link. Chlorine demand was calculated based on a first-order assumption as defined by Eq 5-1.

$$C = C_o e^{-kt} \qquad\qquad \text{(Eq 5-1)}$$

Where:

C = the concentration at time t, in mg/L

C_o = initial of chlorine concentration, in mg/L

k = decay rate, in min^{-1}

t = time, in min

e = base of natural (Napierian) logarithms

A bench study was conducted in which chlorine demand for the raw water was calculated using Cherry Hill/Brushy Plains water. The chlorine decay rate was calculated as 0.55 day^{-1}. This decay rate might be considered as the bulk decay rate or the decay rate of chlorine in the treated water. As can be seen by the ratio in the sixth column of Table 5-2, the total system demand is much higher than just the bulk decay rate. This additional demand is most likely due to pipe wall demand, biofilm, and tubercles, and is very significant. It is conceivable that some system components may ultimately have to be replaced in order to meet water quality goals. The links resulting in dead ends

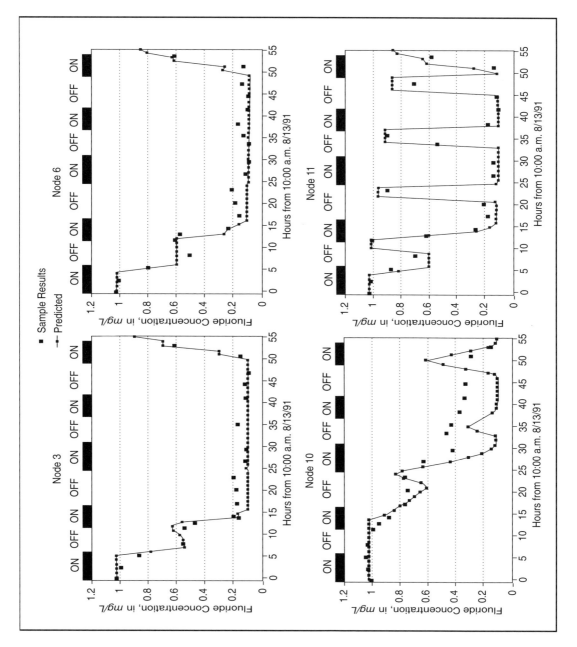

Figure 5-14   Results from fluoride sampling study at nodes 3, 6, 10, and 11

(which also had the lowest average velocity during the "pumps-on" scenario) also exhibited the highest chlorine demand. It is also clear from the fifth column in Table 5-2 that a single first-order decay rate will not predict chlorine residual adequately.

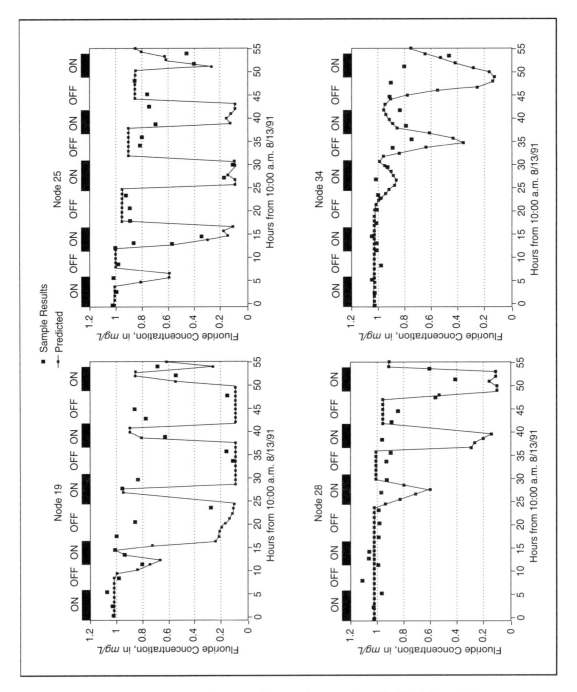

Figure 5-15 Results from fluoride sampling study at nodes 19, 25, 28, and 34

Table 5-2 Hydraulic conditions during "pumps-on" scenario

Link Beginning/ Ending Nodes	Length, ft (m)	Residence Time, days	Chlorine Concentration ratio, (upstream/ downstream)	Decay Coefficient, $days^{-1}$	Ratio of Pipe to Bulk Decay
1–3	3,700 (1,125)	0.0414	1.08/1.00	1.86	3.38
1–6	4,400 (1,341)	0.0321	1.08/1.00	2.40	4.36
6–11	3,800 (1,158)	0.0286	1.00/0.98	0.71	1.29
6–10	4,900 (1,493)	0.8049	1.00/0.36	1.27	2.30
11–19	5,400 (1,645)	0.1634	0.98/0.16	11.09	20.16
11–25	4,350 (1,325)	0.0424	0.98/0.94	0.98	1.78
11–34	64,000 (19,507)	1.3714	0.98/0.12	1.53	2.78
25–28	2,400 (731)	1.4937	0.94/0.16	1.19	2.16
Tank	—	3.000	0.94/0.16	0.59	1.07

Summary

Results from the SCCRWA study were consistent with the findings of others in showing that water quality varies dramatically from point-to-point in a drinking water distribution system (Geldreich et al. 1972; Maul, El-Shaarawi, and Block 1985a and b; LeChevallier, Babcock, and Lee 1987). This study went beyond these earlier studies by presenting a plausible hypothesis as to why this deterioration takes place and demonstrated how system design and specific operating scenarios, such as direct pumping and/or the use of storage tanks, could influence the nature of water quality variation within the system. It also demonstrated the usefulness of water quality propagation models in assessing both the impact and the causative mechanisms for those variations.

References

Clark, R.M., W.M. Grayman, R.M. Males, and A.F. Hess. 1993. Modeling Contaminant Propagation in Drinking Water Distribution Systems. *Jour. Environ. Eng.*, 119(2):349–364.

Clark, R.M., and J.A. Goodrich. 1993. Water Quality Modeling in Distribution Systems. In *Strategies and Technologies for Meeting SDWA Requirements.* Lancaster, Pa.: Technomics Publishing Company.

Clark, R.M., W.M. Grayman, R.M. Males, and J. Coyle. 1988. Modeling Contaminant Propagation in Drinking Water Distribution Systems. *Jour. Water Supply Research and Technology– Aqua*, 3:137–151.

Geldreich, E.E., H.D. Nash, D.J. Reasoner, and R.H. Taylor. 1972. The Necessity of Controlling Bacterial Populations in Potable Water: Community Water Supply. *Jour. AWWA*, 64(a):596–602.

Gessler, J., and T.M. Walski. 1985. *Water Distribution System Optimization*, TREL-85-11, WES. Vicksburg, Miss.: US Army Corps of Engineers.

LeChevallier, M.W., T.M. Babcock, and R.G. Lee. 1987. Examination and Characterization of Distribution System Biofilms. *Applied and Environmental Microbiology*, 53:2,714–2,724.

Maul, A., A.H. El-Shaarawi, and J.C. Block. 1985a. Heterotrophic Bacteria in Water Distribution Systems, I. Spatial and temporal variation. *The Science of the Total Environment*, 44:201–214.

————. 1985b. Heterotrophic Bacteria in Water Distribution Systems, II. Sampling Design for Monitoring. *The Science of the Total Environment*, 44:215–224.

Skov, K.R., A.F. Hess, and D.B. Smith. 1991. Field Sampling Procedures for Calibration of a Water Distribution System Hydraulic Model. In *Water Quality Modeling In Distribution Systems*. Denver, Colo.: American Water Works Association Research Foundation/ USEPA.

Modeling Total Trihalomethanes and Chlorine Decay

Before leaving the treatment plant, drinking water is generally chlorinated as a final disinfection step before being stored in a clearwell or basin. When the water is discharged from the clearwell or basin, it is transported through the distribution system to the consumer. It is presumed that a detectable chlorine residual will minimize the potential for waterborne disease and biofilm growth in the system. Maintenance of chlorine or other disinfectant residuals is generally considered to be a water quality goal in the United States and most US water systems attempt to maintain a detectable residual throughout the distribution system.

As dissolved chlorine travels through the pipes in the network, it reacts with natural organic matter (NOM) in the bulk water, with biofilm and tubercles on the pipe walls, or with the pipe wall material itself. This reaction results in a decrease in chlorine residual and a corresponding increase in disinfection by-products (DBPs), with the amounts depending on the residence time in the network pipes and holding time in storage facilities. Understanding these reactions assists water utility managers in delivering high quality drinking water and in meeting regulatory requirements under the Safe Drinking Water Act (SDWA) and its amendments.

Water quality modeling has the potential for providing insight into the factors that influence the variables affecting changes in distribution systems' water quality. Understanding the factors that influence the formation of total trihalomethanes (TTHMs) and maintenance of chlorine residuals is of particular interest. The modeling program EPANET has proven to be especially useful for modeling both TTHM formation and the propagation and maintenance of chlorine residuals.

One of the first studies to address these issues was conducted by the US Environmental Protection Agency (USEPA) in collaboration with the North

Marin Water District (NMWD) in California (Clark et al. 1994). Another recently completed study conducted jointly by USEPA and the American Water Works Association Research Foundation (AWWARF) examined these same issues (Vasconcelos et al. 1996). In the AWWARF/USEPA study, various types of models were evaluated to describe both of these phenomenon.

Kinetics of Chlorine Decay and TTHM Formation

Within EPANET a mechanism has been provided for considering the loss (or growth) of a substance by reaction as it travels through the distribution system. Reaction is assumed to occur both within the bulk flow and within the pipe-wall material, based on using first-order kinetics. The general expression for chlorine decay in the bulk flow and at the pipe wall is

$$\frac{dc}{dt} = k_b c - \frac{k_f}{r_h}(c - c_w) \qquad \text{(Eq 6-1)}$$

Where:

dc/dt = rate of chlorine decay, in mg/L/d

c = chlorine concentration in the bulk flow, in mg/L

t = time, in days

k_f = mass transfer coefficient, in ft/d

r_h = hydraulic radius

c_w = chlorine concentration at the pipe wall, in mg/L

k_b = bulk phase chlorine decay coefficient, in day^{-1}

The first term in Eq 6-1 is the bulk flow reaction, while the second term includes c_w, which represents the rate at which material is transported between the bulk flow and reaction sites at the pipe wall. If it is assumed that the rate at the wall is first order with respect to c_w and that it proceeds at the same rate as material is transported to the wall so that no accumulation occurs, the following mass balance at the wall holds:

$$k_f(c - c_w) = k_w c_w \qquad \text{(Eq 6-2)}$$

Where:

k_w = wall reaction constant, in ft/d (m/d)

and other variables as defined previously.

Solving for c_w and substituting into Eq 6-1 yields the following reaction rate expression:

$$\frac{dc}{dt} = -k_b c - \frac{k_w k_f c}{r_h(k_w + k_f)} \qquad \text{(Eq 6-3)}$$

$$k_t = k_b + \frac{k_w k_f}{r_h(k_w + k_f)} \qquad \text{(Eq 6-4)}$$

Substituting Eq 6-4 into Eq 6-3 yields

$$\frac{dc}{dt} = -k_t c \qquad \text{(Eq 6-5)}$$

For steady state flow conditions in a single pipe, this rate expression yields a first-order decay model for chlorine, i.e.:

$$c_t = c_0 e^{-k_t t} \qquad \text{(Eq 6-6)}$$

Where:

c_t = concentration of chlorine at any time t, in mg/L

c_0 = initial concentration of chlorine, in mg/L

However, the decay constant in this model is now a function of a bulk decay rate constant, a wall decay rate constant, the molecular diffusivity of chlorine, the water's kinematic viscosity, velocity, and the pipe radius (Clark, Rossman, and Wymer 1995).

With regard to the formation of TTHMs, it has been demonstrated that the rate of formation is a function of time. It has been postulated that DBP formation is governed by the following equation (Clark et al. 1996):

$$DBP = DBP_u - DBP_u^t \qquad \text{(Eq 6-7)}$$

Where:

DBP_u = the ultimate formation potential of an individual DBP

DBP_u^t = the remaining formation potential at time t

We can also make the first-order assumption

$$\frac{dDBP_u^t}{dt} = -k DBP_u^t \qquad \text{(Eq 6-8)}$$

Where:

$$DBP_u^t = DBP_u \text{ at } t = 0$$

Therefore, integrating Eq 6-8 yields

$$DBP = DBP_u(1 - e^{-kt}) \qquad \text{(Eq 6-9)}$$

Equations 6-6 and 6-9 illustrate that both the loss of chlorine residual and the formation of TTHMs are time-dependent and will vary throughout the system depending on propagation conditions. As mentioned earlier, a study to evaluate these changes was conducted by USEPA in the North Marin Water District (NMWD) and is discussed in the following sections.

The North Marin Case Study

The NMWD served a suburban population of 53,000 in or near Novato, Calif. The district used two sources of water—Stafford Lake and the North Marin Aqueduct. The North Marin Aqueduct is a year-round source, but Stafford Lake was used only during the warm summer months when precipitation is virtually nonexistent and when demand is high. Novato, the largest population center in the NMWD service area, is located in a warm inland coastal valley with a mean annual rainfall of 27 in. (685 mm). There is virtually no precipitation during the growing season from May through September. Eighty-five percent of total water use is residential and the service area contains 13,200 single-family detached homes, which accounted for 65 percent of all water use (Clark et al. 1994).

The water quality of the two sources differs greatly. Stafford Lake water has a high humic content and is treated with conventional treatment and pre-chlorination doses of between 5.5 and 6.0 mg/L. The treated water has a residual of 0.5 mg/L when it leaves the treatment plant clearwell. Total trihalomethane formation potential (TTHMFP) levels in the Stafford Lake water were very high. The North Marin Aqueduct water was derived from the Raney Well Field along the Russian River. While technically groundwater, the source water was likely to contain a high proportion of naturally surface-filtered water. Only aqueduct water is disinfected and is very low in precursor material with a correspondingly low TTHMFP. Both sources carry a residual chlorine level of approximately 0.5 mg/L when the water enters the system.

Figure 6-1 is an overall schematic that shows the entire NMWD service area. Figure 6-2 shows the distribution network for zone 1, which was the major focus of the study discussed in this chapter. Figure 6-2 shows both sources, a schematic of the major pipes in the service area distribution system, the major tanks and pumps, and the sampling points used in this study. Major transfers of water out of zone 1 are shown by arrows. Storage tank and pump data for zone 1 is contained in Tables 6-1 and 6-2. As mentioned earlier, depending on the time of year and time of day, water enters the system from either one or both of the sources. The North Marin Aqueduct source operates year-round, 24 hours per day. The Stafford Lake source operated only during the peak demand period from 6:00 a.m. to 10:00 p.m. and generally operated for a period of 16 hours per day. Table 6-3 summarizes NMWD water use characteristics.

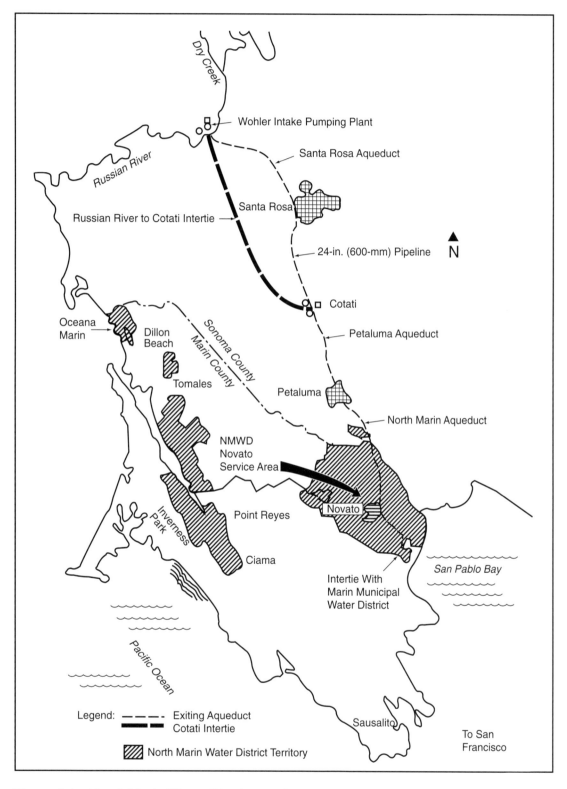

Figure 6-1 North Marin Water District service area

Table 6-1   Zone 1 storage tank area

Tank	Nominal Capacity, gal	Inside Diameter, ft	Bottom Elevation, ft (msl)	Overflow Elevation, ft (msl)	Depth at Overflow, ft
Lynwood #1	500,000	53.0	132.7	163.1	30.4
Lynwood #2	850,000	66.0	131.0	164.4	33.4
Norman	500,000	50.0	123.0	156.8	33.8
Atherton	5,000,000	164.4	133.0	164.5	31.5
Plum*	500,000	60.0	134.0	157.7	22.7
Total Storage	7,350,000				

* Out of service during study

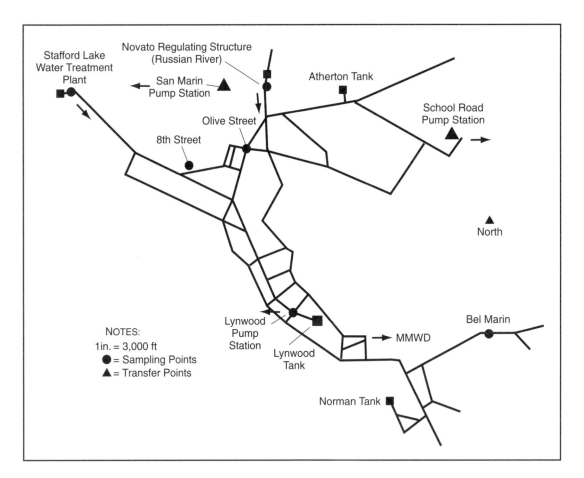

Figure 6-2   Existing zone 1 piping schematic

Table 6-2   Zone 1 to zone 2 pump station data

Pump Station	Pump Number	Rated Horse-power	Rated Flow, gpm	Total Dynamic Head, ft	Rated Revolutions, per minute	Suction Pres-sure, psi	Discharge Pressure, psi	Elevation at Pump, ft	Controlling Reservoir
San Marin	1	100	1,900	165	1,770	47*	118	48	San Mateo
	2	100	1,800	175	1,770	42*	118	48	
	3	75	2,200	100	1,785	75*	118	48	
Lynwood	1	100	1,900	165	1,770	60	131	8	Control from Central on Pacheco Valle Tank Levels
	2	100	1,900	165	1,770	60	131	8	
	3	100	1,900	165	1,770	60	131	8	
School Road	1	30	400	180	3,500	50	130	32	Crest Tank
	2	30	400	180	3,500	50	130	32	
Stafford Lake WTP Service	1	125	3,500	70	1,770	3	42	147	None
	2	125	3,500	70	1,770	3	42	147	
	3	125	3,500	70	NA	3	38	147	

NOTES: 1. Zone 1 to zone 2 pump stations include Hayden pump station (PS), Hancock PS, Diablo Hills PS, and Woodlands PS. They are not included in the analysis since demands are so small in the areas they serve.

2. NA = information not available.

* Suction pressure is a function of SCWA Aqueduct pressure (varies 48–30 psi.)

Table 6-3   North Marin Water District water-use characteristics

Service Area:	100 mi^2 (259 km^2)
Principal Service Center:	Greater Novato area
Population:	53,000
Character:	Suburban (near San Francisco)
Normal Rainfall:	27.2 in. per year (69.1 cm per year)
Normal Reference Evapotranspiration (ET$_o$):	44 in. per year (111.8 cm per year)
Applied Water Requirement for Cool Season Grasses:	27.8 in. per year (70.6 cm per year)
Water Supply:	Surface—North Marin Aqueduct 83 percent, Stafford Lake 17 percent
All Accounts Metered?	Yes
Overall Average Per Capita Use:	135 gpcpd* (511 Lpcpd)
Distribution of Metered Water Use:	Residential (84.7 percent) Commercial (15.3 percent)
Annual Residential Use: Single Family (67.3 percent) Townhouse/Condo (12.7 percent) Mobile Home (16.6 percent) Apartment (3.4 percent)	144 gpcpd (545 Lpcpd) 115 gpcpd (435 Lpcpd) 77 gpcpd (291 Lpcpd) 79 gpcpd (299 Lpcpd)
Unaccounted for Water and Water Loss:	5.7 percent

* Gallons per capita per day (based on average household density of 2.7 persons/dwelling).

EPANET was used to model the system hydraulics, including the relative flow from each source, TTHMs, and chlorine residual propagation (Rossman, Clark, and Grayman 1994). The model was based on an earlier network representation made by Montgomery Watson, Inc., for NMWD and was calibrated based on a comparison of simulated versus actual tank levels for the May 27–29, 1992, period of operation.

Figure 6-3 compares actual flow data from the aqueduct and Stafford Lake against modeled values. The figure shows a discrepancy between the pump as it actually operated during this study period and the model predictions based on tank level elevations. The model was set to operate on Lynwood tank levels, which, in turn, was based on NMWD supervisory control and data acquisition (SCADA) information. However, during the data collection period, as shown in Figure 6-3, the pump was actually shut down five hours before normal operating policy would dictate. This slight change in operation led to a prediction of more water in the system than actually occurred, as shown by Figure 6-4 (Norman and Atherton tanks), although the actual and simulated tank levels for the Lynwood tanks were very close. A second simulation was attempted in which the aqueduct pump was operated on a time schedule. However, the tank level predictions for the Lynwood tank did not match the actual data. It was decided to use the first simulation as the basis for the analysis because the Lynwood tank (Figure 6-4) is the controlling tank for the system. Information from the first simulation, coupled with pump operating schedules and demands in the network, dictated the hydraulic conditions used as the basis for this analysis.

The dynamic nature of the system leads to both variable flow conditions and quality in the network. Flow directions frequently reverse within a given portion of the network during a typical operating day. Figures 6-5 through 6-8 show the changes in flow direction and the percentage of water from Stafford Lake penetrating the system during a typical operating day. In these figures, N5 is the designation for Stafford Lake water. As can be seen, the system experienced dramatic flow reversals, depending on which source or sources

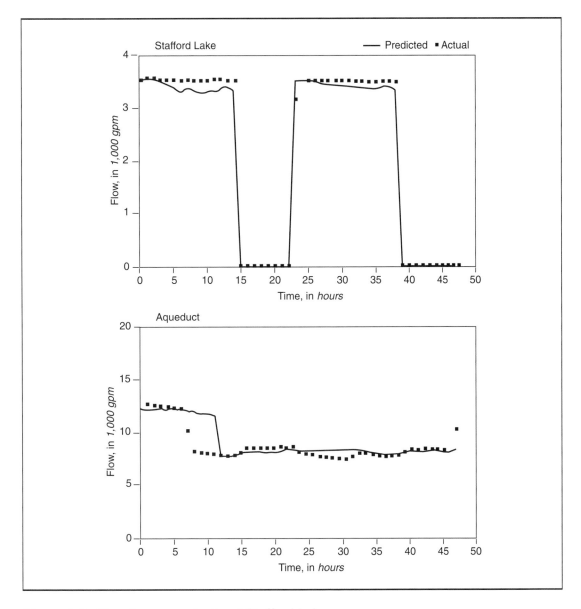

Figure 6-3 Flow from aqueduct and Stafford Lake

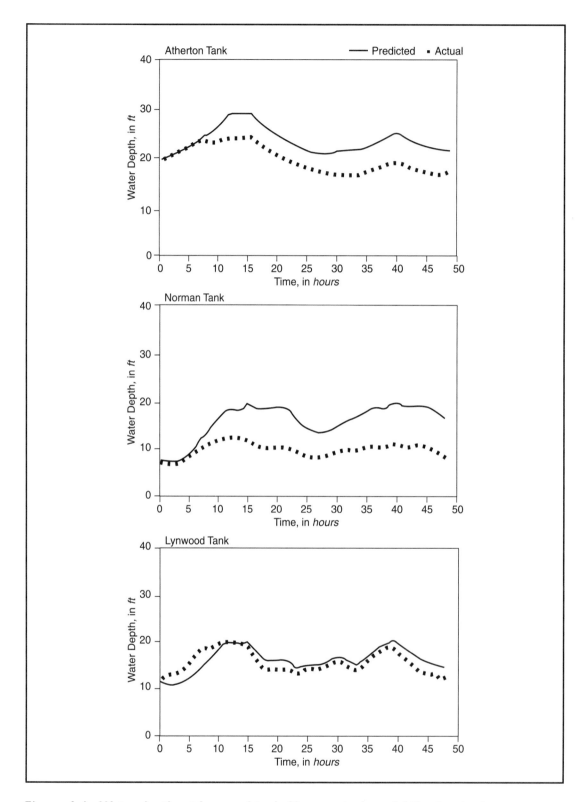

Figure 6-4 Water depths at Lynwood tank, Norman tank, and Atherton tank

were operating. Consequently, the mix of water ranged from 100 percent Stafford Lake water to 100 percent North Marine Aqueduct water at a given node. Figure 6-9 shows the predicted percentage of water from Stafford Lake at the various sampling points used during the two-day study. A discussion of the consequences of these variable flow patterns for water quality follow.

WATER QUALITY STUDY

In order to characterize the NMWD water quality, USEPA designed a sampling protocol and sent a team of investigators to work with NMWD and Montgomery Watson staff during the period May 27–29, 1992. Table 6-4 shows the sampling schedule at six sites. Table 6-5 contains the results of a round-the-clock sampling study conducted over a two-day period. Column 1

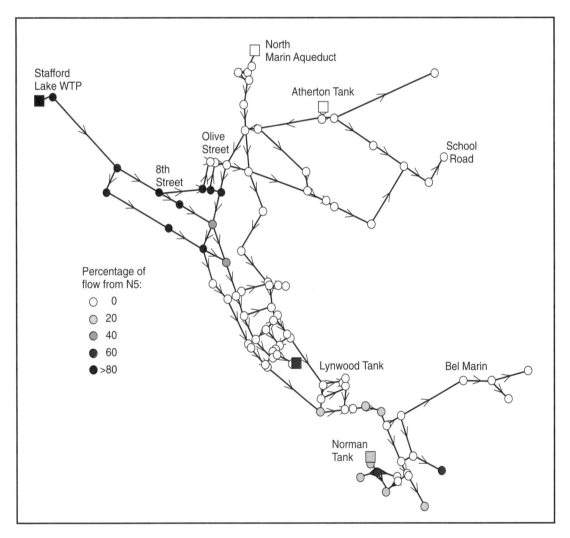

Figure 6-5   North Marin hydraulic calibrations for 6:00 a.m. showing direction of flow and percentage of flow from N5 (Stafford Lake WTP)

of Table 6-5 (page 113) shows the sampling site, date, and time of the sample. Columns 2 through 5 contain values for individual trihalomethane (THM) species. Columns 7 through 10 contain haloacetic acid (HAA) values. Columns 11, 12, 13, 14, and 15 contain values for temperature, free and total chlorine, pH, and sodium, respectively. Figure 6-10 shows the time formation curves for TTHMs and chlorine demand for the two source waters. These figures illustrate the significant difference in water quality from the two sources.

As can be seen from Table 6-5, water quality can change dramatically over time in the system. For example, at the Eighth St. sampling point, trichloromethane ($CHCl_3$) levels vary from 38.4 to 120.1 µg/L over the two-day period. This variability is due to the penetration of water from the two different sources. As can be seen from Figure 6-9, a sample taken at Eighth St. may consist of water from the aqueduct, Stafford Lake, or a blend of both. However, over the sampling period, THM levels were relatively constant at the two

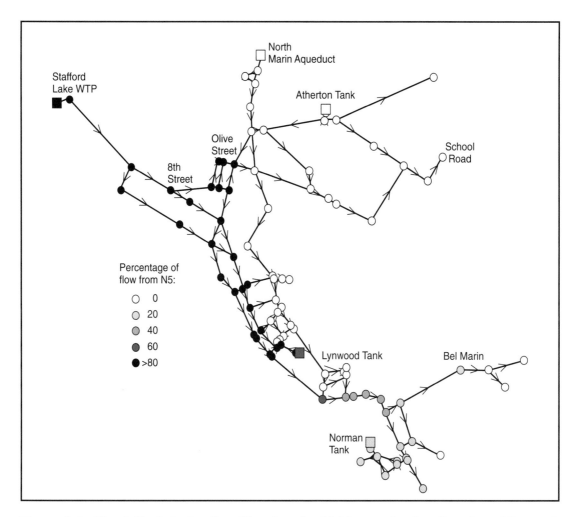

Figure 6-6   North Marin hydraulic calibrations for 12:00 a.m. showing direction of flow and percentage of flow from N5 (Stafford Lake WTP)

sources. Given the extreme differences in water quality from the two sources and the variability in percent source water penetrating within the network, it is reasonable to assume that mixing and blending of water is a major factor affecting water quality.

MODELING TTHMS

In order to test the mixing hypothesis, the average values for TTHMs at each of the sources were used. These TTHM levels were treated as conservative and TTHM levels at each sampling station were calculated based on the percentage of water from a source at that point over time. This assumption was verified with field and bench testing. For example, the average TTHM levels for Stafford Lake water and for the aqueduct were 151.1 and 20.7 µg/L, respectively. If, at a given time and place in the system, 50 percent of the flow is from

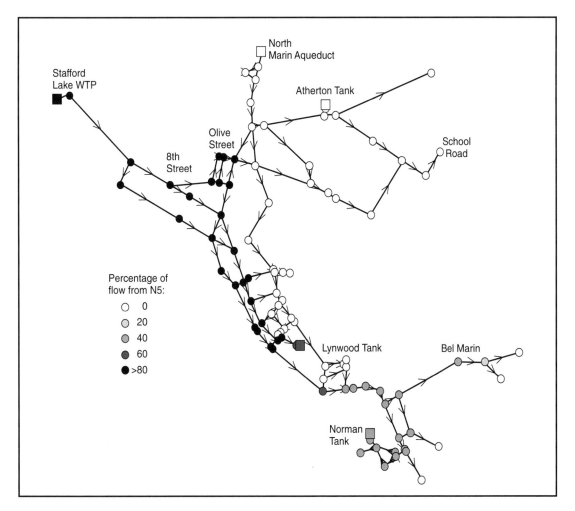

Figure 6-7 North Marin hydraulic calibrations for 6:00 p.m. showing direction of flow and percentage of flow from N5 (Stafford Lake WTP)

Stafford Lake and 50 percent is from the aqueduct, then the predicted TTHM values at that point would be 85.9 µg/L. Figure 6-11 shows the results of this assumption at the various sampling points. Some of the differences between actual and predicted values in the figures are likely because of the calibration problems discussed earlier.

It was observed that ultraviolet (UV) radiation absorbance and TTHMs were closely related. In order to establish that UV absorbance is a good predictor for TTHMs, a regression relationship between UV absorbance and TTHMs was developed using the data from the two sources. Data from the first day of the study was used to develop a regression model between TTHMs and UV as follows:

$$TTHM = 2.056 + 1648.2 * UV \hspace{3cm} (Eq\ 6\text{-}10)$$

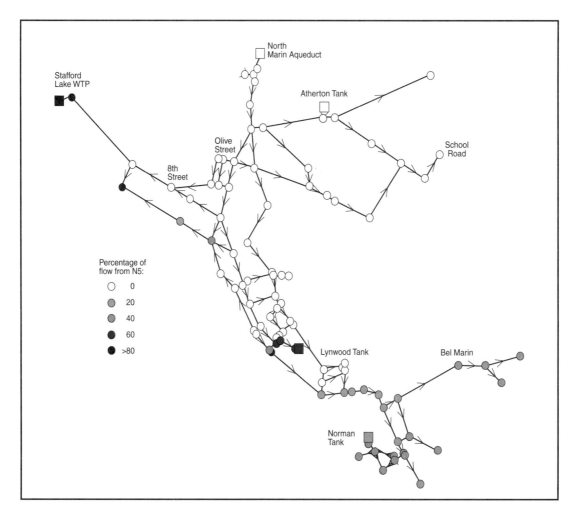

Figure 6-8   North Marin hydraulic calibrations for 12:00 p.m. showing direction of flow and percentage of flow from N5 (Stafford Lake WTP)

Where:

TTHM = total trihalomethanes, in µg/L

UV = ultraviolet absorbance in the 250-nm range, in cm^{-1}

with a p-value = 0.0001, r = 0.9875, and MSE = 11.1. The assumptions of the model (constant variance and normality of error terms) were checked and deemed reasonable. Figure 6-12 shows the data used to develop the regression model. Figure 6-13 shows the use of Eq 6-10 to predict TTHMs at the various sampling points for the second day of the study. As can be seen, the predictions are quite good.

Table 6-4 Sampling schedule

Sampling Times	Locations					
	1	2	3	4	5	6
Wednesday (May 27, 1992)						
6:00 a.m.	X	X	X	X	X	X
8:00 a.m.						X
10:00 a.m.						X
12:00 a.m.	X	X	X	X	X	X
2:00 p.m.						X
4:00 p.m.						X
6:00 p.m.	X	X	X	X	X	X
8:00 p.m.						X
10:00 p.m.						X
12:00 p.m.	X	X	X	X	X	X
Thursday (May 28, 1992)						
2:00 a.m.						X
4:00 a.m.						X
6:00 a.m.	X	X	X	X	X	X
8:00 a.m.						X
10:00 a.m.						X
12:00 a.m.	X	X	X	X	X	X
2:00 p.m.						X
4:00 p.m.						X
6:00 p.m.	X	X	X	X	X	X
8:00 p.m.						X
10:00 p.m.						X
12:00 p.m.	X	X	X	X	X	X

NOTES:
1 = Stafford Lake WTP send-out. 3 = Bel Marin. 5 = 1100 Olive St.
2 = Aqueduct source. 4 = Eighth St. area. 6 = Lynwood tank area.

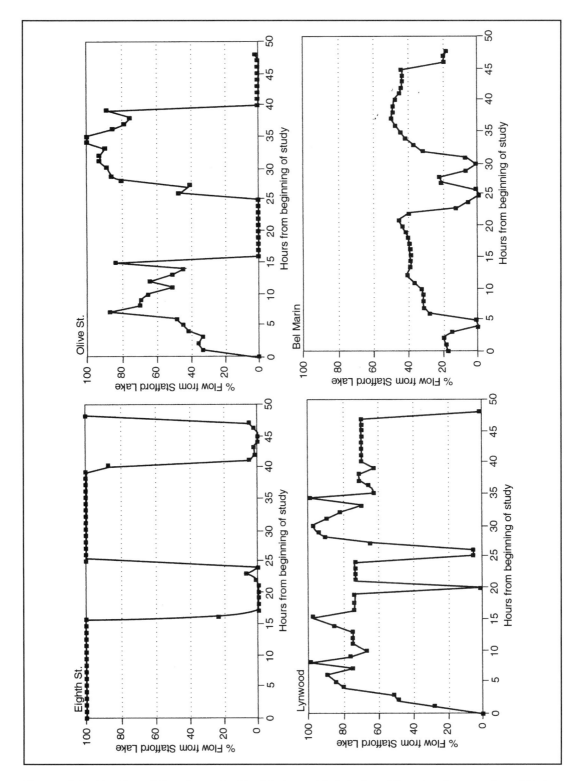

Figure 6-9   Predicted flow from Stafford Lake at Eighth St., Olive St., Lynwood, and Bel Marin sampling sites (percentage)

Table 6-5 Water quality characteristics of NMWD sources and sampling points

Station	CHCl3,* µg/L	CHBrCl2,* µg/L	CHBr2Cl,* µg/L	CHBr3,* µg/L	UV,†	DCAA,‡ ppb	TCAA,‡ ppb	BCAA,‡ ppb	DBAA,‡ ppb	Temp, °F	Cl2-F§	Cl2-T§	pH**	Na,†† µg/L
Aqueduct	0.8–3.0	3.1–6.4	7.0–11.4	4.8–7.5	0.011–0.012	<0.80	<0.78	<1.89–2.27	<0.99–7.08	66–70	0.66–0.45	0.32–0.45	7.40	9.50–10.00
Bel Marin	9.6–81.8	6.3–22.0	6.3–8.6	2.9–4.8	0.019–0.063	<0.80–3.97	1.79–22.63	<1.39–2.07	<0.99	68–73	0.02–0.22	0.12–0.23	—	9.20–20.00
Eighth St.	38.4–129.9	12.4–30.7	2.7–8.3	<0.1–4.7	0.032–0.094	<0.80–43.00	9.33–52.56	<1.89–6.50	<0.99	68–73	0.05–0.61	0.17–0.84	—	12.00–29.50
Lynwood	3.6–124.5	4.2–28.9	4.0–8.4	<0.1–5.6	0.015–0.092	<0.80–19.10	0.78–59.06	<1.89–2.94	<0.99	68–73	0.00–0.55	0.60–0.56	7.50–8.36	9.20–27.00
Olive St.	9.0–134.7	5.2–30.2	3.0–10.3	<0.1–6.1	0.015–0.093	<0.80–36.11	1.37–49.69	<1.89–4.46	<0.99	67–72	0.02–0.41	0.14–0.51	—	9.40–29.00
Stafford WTP	108.0–140.5	25.8–32.1	2.4–3.8	<0.1	0.088–0.095	<2.90–53.32	43.54–61.92	5.65–7.63	<0.99	67–71	0.40–0.58	1.08	8.90–9.10	27.00–30.00

* USEPA (1990), method 551.

† APHA, AWWA, and WEF (1989), method 5910A0.

‡ USEPA (1990), method 552.

§ APHA, AWWA, and WEF (1989), method 409E.

** USEPA (1979), method 150.1.

†† USEPA (1979), method 273.1.

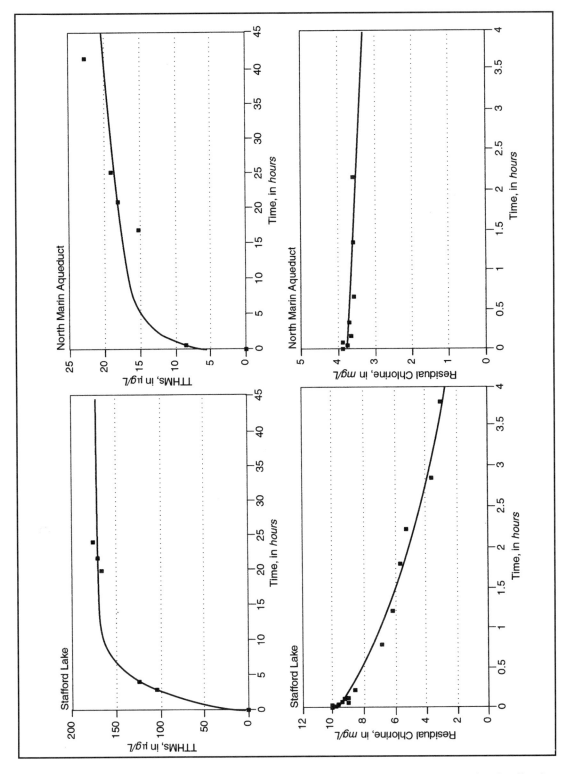

Figure 6-10 TTHM formation potential curves and chlorine demand curves for Stafford Lake and North Marin aqueduct

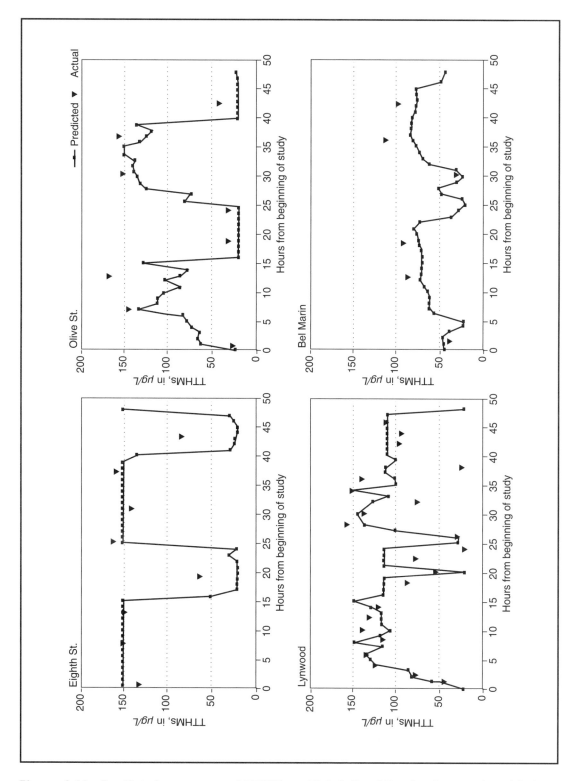

Figure 6-11 Predicted versus actual TTHMs at Eighth St., Olive St., Lynwood, and Bel Marin sampling sites

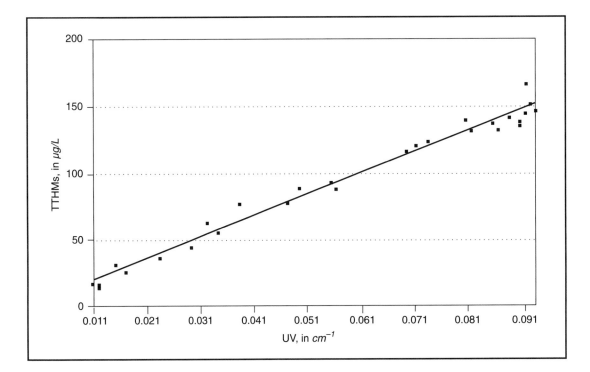

Figure 6-12   Ultraviolet radiation absorbance versus TTHMs from first day of water quality study

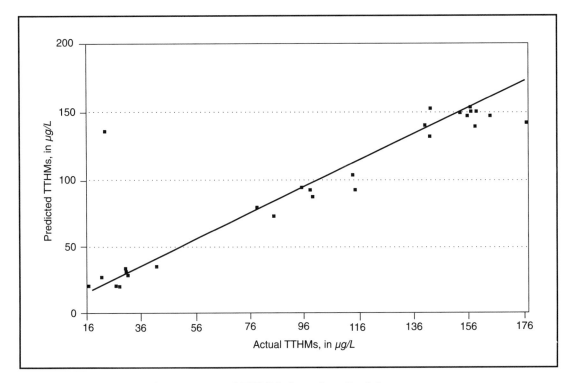

Figure 6-13   Predicted versus actual TTHMs based on Eq 6-1

Figure 6-14 shows simulated versus actual UV absorbance data at the four sampling sites based on the same mixing assumptions used for predicting trihalomethane levels. Eq 6-10 can be used to predict THMs using UV absorbance or the independent variable.

CHLORINE DEMAND

Both Stafford Lake and the North Marin aqueduct water generally maintain a 0.5-mg/L chlorine residual level as the treated water enters the system. Figure 6-10 illustrates chlorine demand for the two sources. As can be seen, the Stafford Lake water has a much higher chlorine demand than does the North Marin aqueduct water. In order to predict chlorine demand at the various sampling points a first-order decay relationship was assumed as follows (Clark et al. 1993):

$$C_t = C_o e^{-kt} \qquad \text{(Eq 6-11)}$$

Where:

C_t = chlorine concentration at any time t, $\mu g/L$

C_o = initial chlorine residual, $\mu g/L$

k = first-order decay coefficient, $hours^{-1}$

t = time, $hours$

e = base of natural (Naperian) logarithms

In EPANET chlorine decay is represented by decay in the bulk phase and pipe-wall decay. Based on bulk water calculations, the first-order decay coefficients or bulk demand for the Stafford Lake and North Marin aqueduct sources were 0.31 day^{-1} and 0.03 day^{-1}, respectively. The chlorine residual was estimated using EPANET and the previously assumed hydraulic conditions. Figure 6-15 shows these predictions based on the four sampling points.

EFFECT OF SYSTEM DEMAND

It is evident that the pipes in the distribution network can exhibit a demand for chlorine. This demand may come from tubercles, biofilm, and perhaps the pipe-wall material itself (Clark et al. 1993). A comparison between chlorine residuals using the first-order assumptions predicted from EPANET versus actual chlorine residuals provides an excellent illustration of this point.

It is clear that there is demand in the system beyond just bulk water decay. Because EPANET has the capacity to incorporate a wall demand factor in addition to the bulk demand factors for chlorine, the system was simulated again using the bulk demand for the two sources and trial and error was used to estimate wall demands for four sections of the network as shown in

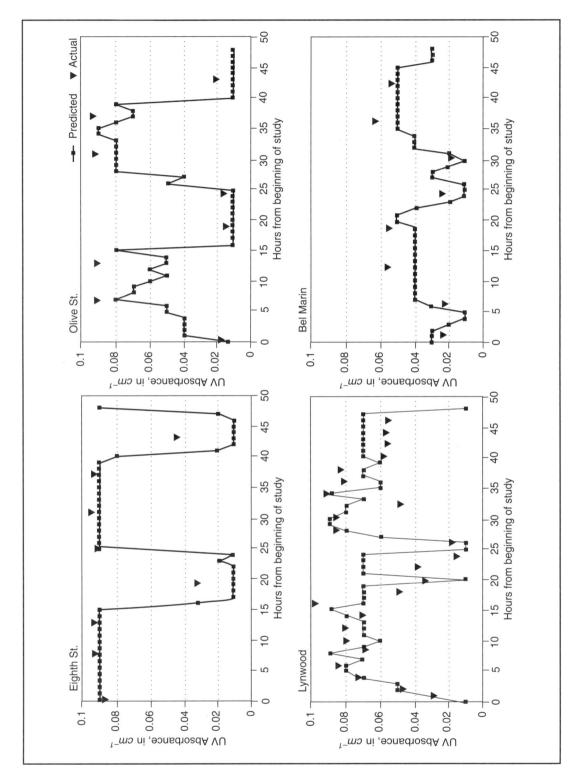

Figure 6-14 Predicted versus actual UV absorbance at Eighth St., Olive St., Lynwood, and Bel Marin sampling sites

Figure 6-16. Wall demand k_w, as defined in EPANET, is given in feet per day. One might think of k_w as a penetration velocity for the disinfectant.

Chlorine residuals were reestimated at the four sampling points (Figure 6-17) and, as can be seen, wall demand obviously plays a major role in chlorine residual loss in the NMWD system. Figure 6-17, however, still shows some large differences between predicted and actual chlorine residuals during the middle portion of the study.

The wall demand effect may be due to the source or to the age of the system. For example, as illustrated by Figure 6-16, the maximum wall demand was found in those areas served primarily by Stafford Lake. However, these pipes are also the oldest in the system.

AWWARF/USEPA Study

As a complement to the North Marin study, USEPA and AWWARF (Vasconcelos et al. 1996) initiated a project that involved several case study utilities and was intended to gain a better understanding of the kinetic relationships describing chlorine decay and TTHM formation in water distribution systems.

Several kinetic models were evaluated, tested, and validated using data collected in these field-sampling studies based on the EPANET distribution network model as the framework for analysis.

Another goal was to provide information and guidelines for conducting water quality sampling and modeling studies by water utilities. Chapter 7 presents some of this information as general guidelines for conducting network modeling studies. The specific objectives of the USEPA/AWWARF study were as follows:

1. Establish protocols for obtaining data to calibrate predictive models for chlorine decay and TTHM formation.

2. Evaluate alternative kinetic models for chlorine decay and TTHM formation in distribution systems.

3. Develop a better understanding of the factors influencing the reactivity of chlorine in distribution and transmission pipes.

4. Make these results available for use in system operations and in planning similar water quality studies.

Utilities selected for participation in this study represent a wide range of source water qualities. In addition, some systems were included that had small isolated zones or long isolated lines. Systems having significant TTHM formation potential were also selected. The utilities in the study are as follows:

- Bellingham, Wash.

- United Water Corporation (formerly Dauphin Water Company), Harrisburg, Pa.

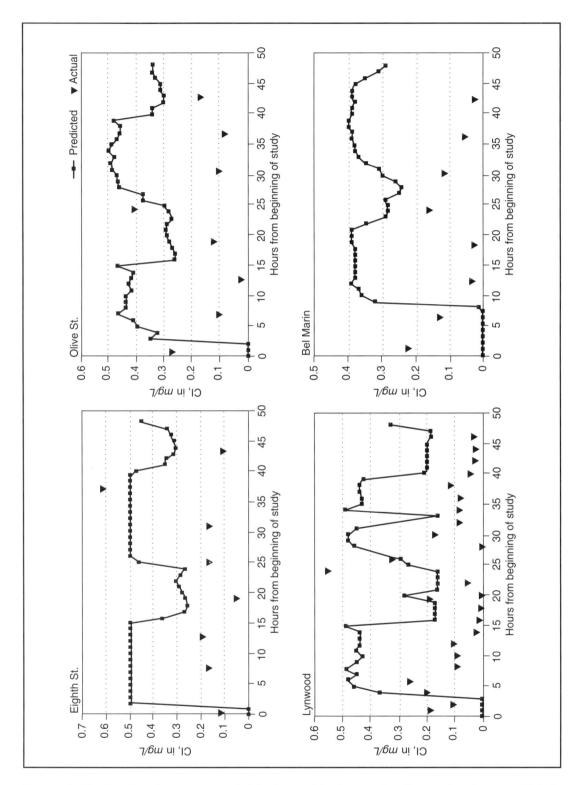

Figure 6-15   Predicted versus actual chlorine residual assuming first-order decay at Eighth St., Olive St., Lynwood, and Bel Marin sampling sites

- Fairfield, Calif.

- North Marin Water District, Novato, Calif.

- North Penn Water Authority, Lansdale, Pa.

- Parisienne des Eaux, France

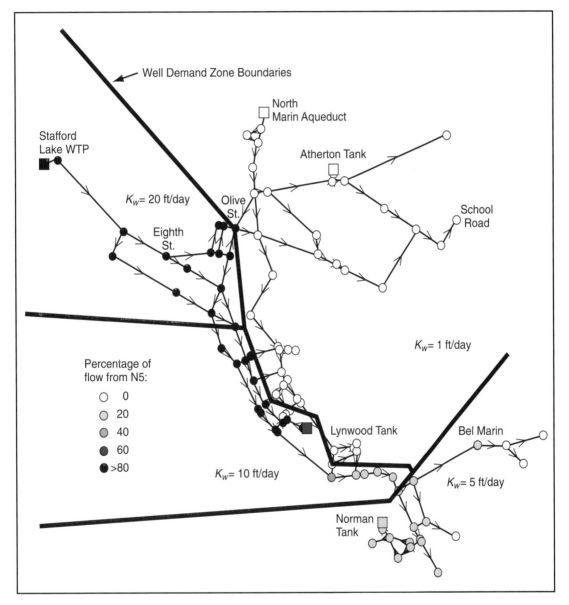

Figure 6-16 North Marin hydraulic calibration for 12:00 a.m. incorporating wall demand by zones

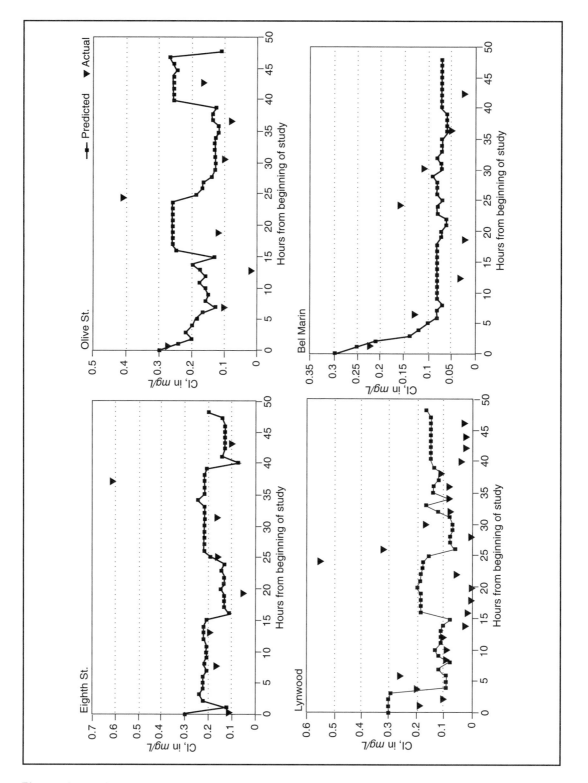

Figure 6-17 Predicted versus actual chlorine residual assuming wall demand factors at Eighth St., Olive St., Lynwood, and Bel Marin sampling sites

FIELD SAMPLING STUDIES

Extensive water quality sampling was conducted at the five participating US utilities. Information was collected on chlorine residuals, TTHM concentrations, and other relevant water quality and hydraulic factors. These data were used to calibrate a hydraulic model and a water quality network model for each test network and to calculate chlorine decay and TTHM formation kinetic constants. Tracer studies were performed to aid in refining the hydraulic calibration for four of the five utilities and in each utility, a subnetwork and/or a long pipeline was selected for sampling.

Sampling plans were prepared that contained detailed descriptions of all aspects of each study, including contingency plans addressing how to deal with unexpected occurrences. A final report was prepared containing all aspects of the field work at the end of each individual study (Vasconcelos et al. 1996).

Evaluation and Testing of Alternative Kinetic Models

The following chlorine decay models were evaluated: (1) first order, (2) *nth* order, (3) limited first order, and (4) parallel first order for bulk decay (Table 6-6). First-order or zero-order decay kinetics were considered for wall decay. Results from this analysis suggested that for TTHM data, limited first-order growth kinetics might be used as the basis for predicting TTHM formation in distribution systems.

Testing Chlorine Kinetic Models. A method similar to that described in the North Marin study was developed for systematically analyzing alternative kinetic models. After determining if wall decay was significant, a series of simulations was performed using alternative wall decay kinetic models in which the wall reaction rate coefficient was systematically adjusted until the best statistical comparison between the observed and simulated chlorine concentrations was obtained. This procedure was repeated for each wall reaction kinetic model tested.

Table 6-6   Mathematical models for alternative decay models

Model	Differential Form ($dC/dt =$)
First order	$-kC$
nth order	$-kC^n$
Limited first order	$-k(C - C^*)$
Parallel first order	$-k_1 C_1 - k_2 C_2$ with $C_{1,0} = C_0 x$ and $C_{2,0} = C_0(1 - x)$

NOTES:

C = concentration
C_0 = initial concentration
t = time
$k, n,$ and C^* = parameters to be estimated

THM Kinetic Model.   Trihalomethanes were modeled using a first-order growth, limited reaction-rate model, in which the THM formation potential of the source water was assumed to be the limiting concentration that THMs can reach. This simplified model approach is valid only for a system with a single source water quality.

Research on the Effect of Water Quality Parameters on Chlorine Decay

Lyonnaise des Eaux performed a series of studies on the effects of various water quality parameters and other factors on the kinetics of chlorine decay. The following findings were obtained:

- Total organic carbon (TOC) and temperature influence the rate of chlorine decay.

- The chlorine concentration level is a factor influencing the rate of chlorine decay throughout the distribution network; the lower the chlorine concentration, the higher the rate of chlorine consumption.

- For the reaction $2Fe^{+2} + HOCl \rightarrow 2Fe^{+3} + Cl^- + OH^-$ in the aqueous phase, even in the presence of oxygen, free chlorine, as well as $HOCl$ or ClO^-, is the chemical specie that reacts preferentially with ferrous iron.

Experimental results showed that for this range of conditions, the reaction of free chlorine with ferrous iron is rapid and total. In all cases, reaction speed is high; so the reduction–oxidation (redox) reaction will not be a limiting factor in chlorine consumption due to corrosion.

In distribution networks, biofilm forms on the surface of metal pipes or plastic pipes. In metal pipes, when combined with the corrosion phenomenon, the biofilm is an important factor in chlorine consumption.

Conclusions

A number of conclusions can be drawn from the AWWARF/USEPA study and from the other studies conducted by USEPA. These are as follows:

- Chlorine decay in distribution systems can occur because of reactions in the bulk phase and at the pipe wall.

- Chlorine decay bottle tests of a variety of US waters showed that *nth* order or parallel first-order kinetic models provided better fits to the data than first-order models, although in most cases, the difference was minimal.

- Laboratory studies showed that pipe-wall reactions related to corrosion of ferrous pipe materials can consume significantly more chlorine than those related to biofilm.

- The rate of reaction of chlorine at the pipe wall is inversely related to pipe diameter and can be limited by the rate of mass transfer of chlorine to the wall.

- At this time there is no available method for directly determining the kinetics of chlorine decay due to pipe-wall reactions. Field data must be used to calculate these reactions.

- A well-calibrated hydraulic model, one preferably based on a tracer study, is an absolute requirement for a distribution system water quality study.

- The first-order TTHM formation model showed acceptable performance in limited testing in actual distribution systems.

- The enhanced version of the EPANET software developed for this project is an effective tool for calibrating network models for chlorine decay and TTHM formation.

References

American Public Health Association, American Water Works Association, and Water Environment Federation. 1989. *Standard Methods for the Examination of Water and Wastewater*, 17th ed. Supplement 1. Washington, D.C.: American Public Health Association.

Clark, R.M., J.A. Goodrich, and L.J. Wymer. 1993. Effect of the Distribution System on Drinking Water Quality. *Jour. Water Supply Research and Technology–Aqua*, 42(1):30–38.

Clark, R.M., G. Smalley, J.A. Goodrich, R. Tull, L.A. Rossman, J.T. Vasconcelos, and P.F. Boulos. 1994. Managing Water Quality in Distribution Systems: Simulating TTHM and Chlorine Residual Propagation. *Jour. Water Supply Research and Technology–Aqua*, 43(4):182–191.

Clark, R.M., L.A. Rossman, and L.G. Wymer. 1995. Modeling Distribution System Water Quality: Regulatory Implications. *Jour. Water Resources Planning and Management*, 121(6):423–428.

Clark, R.M., H. Pourmoghaddas, L.G. Wymer, and R.C. Dressman. 1996. Modelling the Kinetics of Chlorination By-product Formation: The Effects of Bromide. *Jour. Water Supply Research and Technology–Aqua*, 45(1):1–8.

Rossman, L.A., R.M. Clark, and W.M. Grayman. 1994. Modeling Chlorine Residuals In Drinking Water Distribution Systems. *Jour. Environ. Eng.*, 120(4):803–820.

USEPA. 1990. *Methods for the Determination of Organic Compounds in Drinking Water*, Supplement 1. EPA/600/4-90/020.

———. 1979. *Methods for the Analysis of Water and Wastewater*, EPA/600/4-79-020.

Vasconcelos, J.J., P.F. Boulos, W.M. Grayman, L. Kiene, O. Wable, P. Biswas, A. Bhari, L.A. Rossman, R.M. Clark, and J.A. Goodrich. 1996. *Characterization and Modeling of Chlorine Decay in Distribution Systems*. Denver, Colo.: American Water Works Association Research Foundation and American Water Works Association.

Applying
Water Quality Models

Applying water quality models to distribution systems requires careful attention to a number of details. This chapter discusses some of the experiences gained from studies conducted by the US Environmental Protection Agency (USEPA) (Clark et al. 1994a and b) and the American Water Works Association Research Foundation (AWWARF) (Vasconcelos et al. 1996).

Three important elements involved in applying water quality models effectively are

- an understanding of the effect that network configuration has on propagation of substances and contaminants

- sampling and analysis procedures

- design of the field sampling program

One of the initial steps in applying a water quality model to a given system is to establish a database for the service area. First, the pipe network is defined. This involves acquiring information on water usage and node elevations, and establishing a link–node file that contains various types of information required to model and display the distribution system. The size and configuration of the network is also important in designing a field study.

Three general network configurations that might be studied include the following: isolated pipes; small and isolated portions of a distribution system; and larger pipes in all, or major portions, of a large distribution system (as illustrated in Figure 7-1). No matter the size of the system, the objective is to collect sufficient information so that a model can be calibrated.

If a single pipe is to be studied, it may be viewed as a highly controlled experiment. A pipe of sufficient length should be selected so that variables such as chlorine decay can be measured and so that pipe velocities and characteristics (e.g., diameter, material, and conditions) are known and remain constant over a length of the pipe. Pipes of small diameter (i.e., <12 in. [305 mm])

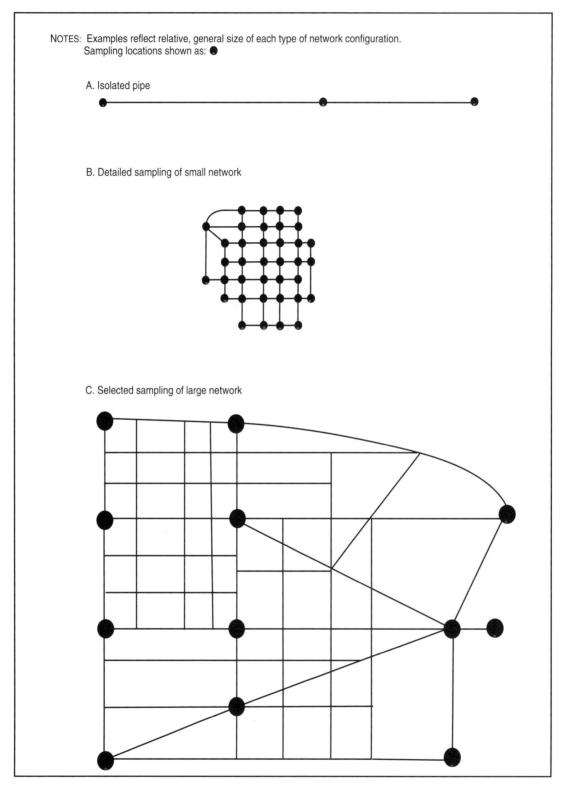

NOTES: Examples reflect relative, general size of each type of network configuration.
Sampling locations shown as: ●

A. Isolated pipe

B. Detailed sampling of small network

C. Selected sampling of large network

Figure 7-1   Examples of alternative network configurations

would be desirable if, for example, the objective is to assess the impact of wall demand. Measurements must be taken at the upstream and downstream end of the pipe (and, if possible, at intermediate points) and flow rates and pipe diameter must be known in order to determine travel times. By studying these pipes under different flow conditions over several seasons, site-specific changes in such variables as total trihalomethane (TTHM) formation or chlorine decay may be determined.

Frequently a type of configuration to be considered is a small network or an isolated portion of a larger network where detailed sampling is performed at a large number of locations. Understanding temporal and spatial variations in water usage is very important in understanding the operation of such systems. Therefore, special care must be taken to ensure that good estimates for water usage are available throughout the study area. If sufficient data are available, model variables may be adjusted to reflect field conditions and possibly to develop a relationship between pipe characteristics (such as diameter and material) and decay and/or formation rates for the variable being studied. The level of expertise required for working with a database derived from this network configuration is greater than with an isolated pipe.

Still another configuration to be studied is a large network in which sampling is performed at selected sites within the system. The difference between this configuration and the small network configuration is that the ratio of sampling points to the number of pipes in the network is quite small for the large network study. These data are also useful in adjusting or validating previously developed models.

Overview of Sampling and Analysis Procedures

A general overall approach to designing sampling programs has evolved during the course of the studies described in this book. The effort involves four major steps: (1) development of a detailed sampling plan, (2) performance of the field sampling study, (3) postsampling data analysis and preparation of a report documenting the field study, and (4) analysis of the field data and development (parameterization) of the network model. A brief overview of each of these steps is provided below and discussed in greater detail later in this chapter.

Perhaps the most critical step in performing a field study is the development of a sampling plan. The plan should reflect the specific needs and goals of the study and reflect workforce and resource requirements and constraints. Generally, such a plan should err on the side of too much detail, but at the same time provide sufficient flexibility to allow for the unexpected events that always occur in field work. A written sampling plan should be developed and reviewed by appropriate personnel.

Planning and Preparing for a Sampling Study

The sampling plan serves as a detailed, written description as to how the sampling study is to be performed. It contains a detailed description of all aspects of the study, including contingency plans addressing unexpected occurrences.

As part of the sampling plan, the network hydraulic–water quality model may be applied to the sampling area using the operating conditions likely to be in effect during the field study. Based on predicted results, sampling locations and sampling frequency can be established. In some cases, alternative operating conditions are simulated in order to determine their effect on the planned sampling strategy. Even though the model may not be fully calibrated, and in fact, may only be an approximation of the actual system, the information gained by predicting the behavior of the system under the expected sampling conditions is frequently helpful. Sampling plans should address the specific issues discussed in the following sections.

SAMPLING LOCATIONS

The actual sampling study should follow the sampling plan as closely as possible. If possible, field data should be examined frequently during the course of the study in order to identify possible problems and to determine deviations from the anticipated results that may require modifications in the plan. Special care should be taken to clearly record all data and to document all aspects of the study.

Specific sampling sites should be carefully and accurately delineated in the plan. Typically, samples can be taken from permanent sampling taps, hydrants, public buildings, businesses, residences, or water utility facilities. For all types of sampling sites, the taps should be allowed to run continuously or care should be taken to be sure that they are adequately flushed prior to each sample to make sure that water is being taken from the main. Each type of sampling location has advantages and disadvantages, as outlined below.

Permanent sampling taps generally provide the best sampling method because they are designed to minimize flushing time, provide access at all times, and to lessen the chances of sample contamination. However, such devices are relatively expensive to purchase and install and would not be cost-effective for a one-time use. Hydrants provide good accessibility and are widely available. Prior to use, they should be flushed well and a sampling tap should be installed. Maps and records that describe the system should be consulted prior to use to determine which main the hydrant is connected to, and the length and diameter of the connector. Travel times from the main to the hydrant should be calculated based on flow rates, estimated distance, and diameter to determine required flushing time prior to each sample.

Other water-utility-owned facilities can also serve as good sampling locations because of 24-hour accessibility. Public buildings (especially, fire houses) are also good candidates because of their general accessibility. Residences and businesses should be considered only as a "last resort" because of limited accessibility, liability, and inconvenience to the water customer.

SAMPLING FREQUENCY

Frequency of sampling should be specified in the sampling plan. If samples are to be taken manually, then a "circuit" is generally established in which samples are taken sequentially by a sampling crew. The circuit should be clearly marked on a map. The time required to sample at a single station and over the entire circuit can be estimated by a preliminary test of the circuit or by estimating the times associated with each aspect of the task (e.g., time to flush the sampling tap, time to take and analyze a sample, time to travel between sites, etc.). Adequate rest time (e.g., bathroom stops and meal breaks) or availability of relief crews should be added to sampling time estimates. Additionally, the plan should include contingency plans if the sampling crews fall behind their anticipated schedule. These plans could include the realignment of the circuits, addition of more crew members, or acceptance of less-frequent sample collection.

An alternative to, or variation on, the circuit approach involves sampling along the tracer front. Using this approach, intensive sampling is initiated during the transition period when the tracer is first approaching a sampling station. This approach is logistically more difficult and requires essentially instantaneous analysis of the data so that decisions to move to the next set of stations can be intelligently made.

SYSTEM OPERATION

The operation of a water system can significantly affect the movement of a tracer through the system. During a sampling study, system operating conditions should be viewed as a controlled set of variables rather than a random set of occurrences. Generally, the system operation should either reflect normal operating conditions or be set to represent a desired condition. For example, one may wish to study the system under tank-filling conditions so as to eliminate the effects of old water being discharged to the distribution system. For either normal operation or a special condition, it may be necessary to precondition the system prior to the start of sampling so that the system may operate in the desired fashion. For example, it may be necessary to operate the system so that the water levels in selected tanks are at particular levels in order to ensure that the system can be operated in the desired manner during the study. Additionally, it may be necessary to select a particular season to perform the study in order to allow desired operational flexibility. For example, in many cases, water systems are "stressed" during hot summer months so that modifications in operation are difficult.

PREPARATION OF SAMPLING SITES

Prior to actual sampling, the site should be adequately prepared. Preparation may include testing of faucets in hydrants, calculation of required flushing time, notification of owners, and marking the sites for easy identification.

SAMPLE COLLECTION PROCEDURES

Procedures for collecting samples should be specified in the plan. These procedures may include required flushing times, methods for filling and marking sample containers, listing reagents or preservatives to be added to selected samples, methods for storing samples, and data logging procedures.

ANALYSIS PROCEDURES

Samples taken in the field may be analyzed at the sampling site, at a field laboratory located in the sampling area, or in a centralized laboratory. Specific procedures to be followed for each type of analysis should be specified in the plan.

PERSONNEL ORGANIZATION AND SCHEDULE

An important part of the sampling plan is a detailed personnel schedule for the sampling study. Ideally, the schedule should include crew assignments and a work schedule for each study participant. Logistical arrangements include lodging for nonresident participants, provision for meals, transportation, etc. A possible source of assistance during the summer would be college students.

SAFETY ISSUES

There are many safety issues associated with sampling studies. These are intensified by round-the-clock sampling and working in unfamiliar areas. The following safety-related concerns should be addressed in the sampling plan: notification of police and other governmental agencies; public notification (e.g., using newspapers or television stations); notification of customers who may be directly affected; issuance of safety equipment, such as flashlights and vests; use of marked vehicles and uniforms identifying the participants as official water utility employees or contractors; issuance of official ID cards or letters explaining their participation in the study; and use of more than one sampler, especially during nighttime or in dangerous areas.

DATA RECORDING

An organized method for recording all data should be devised. A data recording sheet, such as the one shown in Figure 7-2, should be designed and included as part of the sampling plan. Use of military (24-hour) time is

Study: Oberlin System, Dauphin   **Sheet Number:** _____

Date MMDD	Station ID	Time HHMM	Initials	Temp °C	Chlorine Free	Tot		Sample		FL	Comments

Figure 7-2   Study sampling form

recommended. Data should be recorded neatly in ink and the individual data sheets numbered sequentially. Completed data sheets should not be kept in the field in order to reduce the chance of loss. Once the data sheets are centrally assembled, they should be copied and the copies kept separately from the originals. Frequently, a sample will be taken and partly analyzed in the field (i.e., for temperature and chlorine residual) and then stored for later analysis (i.e., for fluoride and THMs). Therefore, a method for relating the field and lab samples is needed. Use of date-time-location on both the recording sheet and on the sample container provides a mechanism for identifying samples.

EQUIPMENT AND SUPPLY NEEDS

Equipment includes field sampling equipment (e.g., a chlorine meter), safety equipment (vests, rain gear, and flashlights), laboratory equipment, etc. Expendable supplies include sampling containers, reagents, and marking pens. As part of the sampling plan, the needs and availability of equipment and supplies should be identified and alternative sources for equipment investigated. Some redundancy in equipment should be planned for, since equipment malfunction or loss is possible.

TRAINING REQUIREMENTS

Training of sampling crews is essential and should be specified in the sampling plan. Even though crews may be composed of experienced workers, it is important that all crews be trained in a consistent set of procedures. Training topics include sampling locations, sample collection and analysis procedures, data recording procedures, contingencies, etc. All crew members should read the sampling plan before the training sessions.

CONTINGENCY PLANS

The old adage: "If something can go wrong, then it will," applies to field sampling studies and provides a good basis for contingency planning. Contingency planning should include equipment malfunction, illness of crew members, communication problems, severe weather, unexpected system operation, and customer complaints. All of these events occurred once or more during the sampling studies discussed in this book.

COMMUNICATIONS

During a sampling study, key personnel are often distributed throughout the area as follows: samplers at single stations or riding a circuit, laboratory analysts in a permanent or field lab, operations personnel at central control stations, and the study supervisor at any of many locations. In order to coordinate actions during the study and/or to respond to unexpected events, some means of communication is needed. Alternatives include radios (normally available if you are using utility vehicles), cellular phones, walkie-talkies, or, the low-tech solution: a person circulating in a vehicle.

CALIBRATION AND REVIEW OF ANALYTICAL INSTRUMENTS

A specific plan for reviewing and calibrating all instruments is necessary and this work must be carried out prior to the study. When multiple instruments are used to measure the same parameter (e.g., multiple chlorine meters), each instrument should be calibrated against a standard and compared. A plan to check the calibration of instruments during the study is also essential.

Conducting the Sampling Study

Ideally, conducting a sampling study would simply involve following the detailed sampling plan to completion. In practice, however, unexpected events occur and adjustments must be made in order to meet the study objectives.

The most successful studies are well planned and executed in an orderly manner. Ideally, a study should be coordinated through a central location where a study supervisor is located. The supervisor should function as the central coordinator, receiving information from the field, lab, and operations center and then making decisions and disseminating information back to the field. Decisions on changes in sampling schedules or operating procedures should be made from the operating center based on full knowledge of all aspects of the ongoing study. Use of a computerized database system or a manual system for plotting the results of the ongoing study can serve the useful role of showing the behavior of the system and identifying variations to the sampling plan.

During a study, it is essential that accurate and complete records be kept at all times. Reliance on memory as to when certain events occurred can lead to inaccurate or incomplete information.

POSTSAMPLING LAB ANALYSIS AND PREPARATION OF DATA REPORT

During and immediately following the completion of the field work, lab analysis should be completed. When analysis permits, samples can be preserved so that if the resulting data appears inconsistent, tests can be repeated. This is especially true of the fluoride samples, which have a relatively long shelf life.

An important step following the field work is the preparation of a data report. The purpose of the data report is to record all pertinent information from the field study and subsequent lab analysis. This includes information on the sampling study, the study area, and results of the lab analysis organized in a usable manner.

All field sampling water quality data should be stored in a computerized database management system (DBMS) in order to facilitate the preparation of a data report and its subsequent use. For studies described in this book, the R:Base DBMS was used with the data organized in a relational structure. The data structure is shown graphically in Figure 7-3. In this relational structure, sampling data are uniquely identified by date, time, and sampling station. The primary table holds the key data describing the location, date, time, and sampler, along with the data that is routinely collected and analyzed in the field (chlorine, fluoride, temperature, etc.). Additionally, a unique identifier is

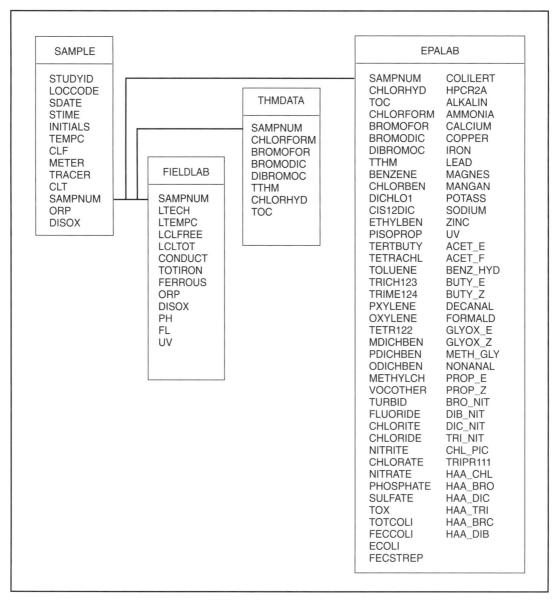

Figure 7-3 Relational database structure for water quality sampling data

assigned to each sample that undergoes additional laboratory analysis. Using the sample identifier as the link to other tables, the additional analysis is stored in tables corresponding to data analyzed in field laboratories.

Analysis of Data and Development of Model

As soon as possible following completion of the field study, laboratory analysis should be completed and a report prepared containing all aspects of the field work, including all details of the study and the results of all field and laboratory analyses. This report serves as the repository of all information on the field study and, as such, the basis for later analysis.

The final step in the process is the actual analysis of field results and use of the data to develop a water quality model. The results of the analysis might include development of a new model, development of a site-specific relationship between the parameters of an existing model and observed pipe characteristics, or verification of an existing set of parameters or relationships.

A two-step method for performing the analysis, including hydraulic calibration of the network model and development of the water quality model, can be used. Because of the relative sensitivity of some variables to travel times, it is highly desirable to have a model that has been hydraulically calibrated for the particular sampling study. As discussed earlier, the use of tracers in conjunction with more traditional calibration techniques is preferred because it should provide the most dependable travel time results.

The methods used to adjust model parameters in order to achieve an acceptable match between field tracer data and model results may vary from informal "sensitivity" methods to more formal search methods. Various methods have proven successful in the studies described in this book. Measures of success have included informal visual matches (e.g., "the predicted model results look like they match field results relatively well") to formal numerical measures of goodness of fit. Two formal goodness of fit measures are illustrated in Figure 7-4. The standard error of estimate is a measure of the deviation of predicted from observed results. The "fuzzy" time measure, developed in this study, recognizes that small temporal variations may be acceptable and adjusts for such small temporal shifts prior to the calculation of the standard error of estimate.

In adjusting parameters to achieve an improved match between observed and predicted results, several general rules are followed:

1. Parameters should be kept within reasonable ranges. For example, roughness coefficients outside of a range from 50 to 160 should be avoided.

2. Adjustment of parameters that have little influence on prediction results is unproductive.

3. Parameters should be adjusted, starting with those for which the greatest degree of uncertainty may be expected, and progressing, if needed, to parameters that are assumed to be known with greater certainty. The following list of model parameters is given in an order that generally approximates increasing levels of certainty. Since there is obviously some subjectivity in defining such an order, and variation between sites would be expected, this list should only be viewed as a general guideline.

A. **Pipe roughness.** In most applications, pipe roughness is derived from information on pipe material and age. The uncertainty in the estimate of pipe roughness is generally improved when in-situ flow tests have been performed.

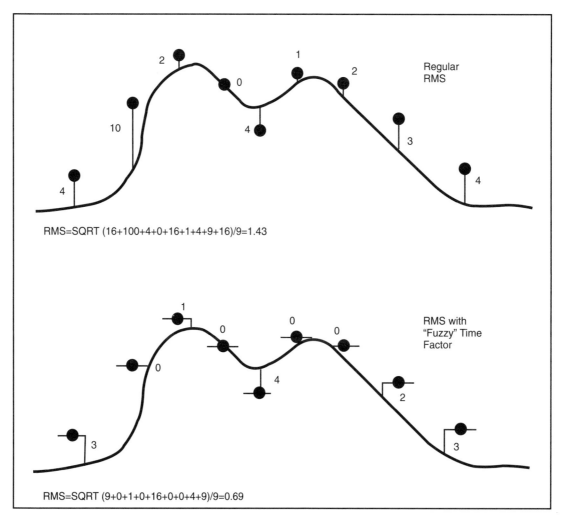

RMS=SQRT (16+100+4+0+16+1+4+9+16)/9=1.43

RMS=SQRT (9+0+1+0+16+0+0+4+9)/9=0.69

Figure 7-4   RMS error calculation

B. **Temporal water usage patterns.** Most dynamic or extended period stimulation (EPS) models require estimates of average (baseline) water use at each node and then a representation of the temporal water use pattern around the baseline value. Continuous meters are typically unavailable at the point of use, thus water use patterns are estimated based on logic (e.g., schools are in session from 8:00 a.m. until 4:00 p.m., so the heaviest usage is during that period) and/or adjusted so that total water usage by time in the system or subsystem is equal to a known pattern, such as the sendout from a plant.

It should be pointed out that as the model representation becomes more detailed to the point that virtually all pipes are included, the uncertainty in temporal patterns at the node level increases because of the stochastic nature of water use at the point of consumption. A possible approach to obtaining demand information would be to use data loggers on a sample of customers to get continuous demand information.

C. **Absolute spatial water usage.** Average water use by node is generally estimated using actual meter readings or is based on land-use or dwelling-type estimates. There is usually significant uncertainty in spatial water use that may be adjusted in order to improve predictions. This is especially true in unmetered systems.

D. **Pump curves.** The relationship between head and flow through pumps is usually available. However, these curves may reflect the pump condition as installed and can be in error. In-situ pump tests reduce any uncertainty associated with these curves.

E. **Valve settings.** There is uncertainty with regard to specific valve parameters, such as critical pressure settings for pressure-reducing valves (PRVs). However, more importantly, some valves may not reflect the expected conditions (i.e., valves that are closed or partially closed when they are expected to be open and vice versa). Such occurrences can cause extreme difficulties in calibrating a model and the user should be ready to try alternative valve settings and to have field crews check existing valve settings.

F. **Other system component data.** Other system component data include flows through meters, water level elevations in tanks, etc. Due to malfunctioning or uncalibrated meters, this information, which is routinely considered as correct, may require checking and/or adjustment in order to improve the calibration.

G. **Tracer decay characteristics.** A tracer should be as conservative (does not degrade in the system) as possible. However, various chemical and physical reactions can result in the loss of an expected conservative substance. In the AWWARF/USEPA project, the loss of a fluoride tracer in a system that did not normally fluoridate was observed and substantiated in the laboratory.

H. **Pipe diameter.** Most modelers view pipe diameter as a known value that does not need to be modified. However, errors on base maps or encrustation of a pipe can result in significant difficulties in calibrating a model. Examination of pipes that have been replaced can indicate whether or not encrustation is an issue that must be addressed.

I. **Tracer concentration measurements.** For systems in which the tracer to be used is not routinely injected, temporary feeding equipment is usually employed. Because such equipment may not function perfectly or the resulting concentrations may vary with flow, frequent concentration measurements should be taken downstream of the injection point and the accuracy of the tracer measurements should be checked at this point and in the system. If there is uncertainty in the measured concentrations, then this parameter may need to be adjusted in the model.

J. **Pipe configuration.** Pipe configuration is generally derived from system maps. The primary uncertainty in this case is unknown closed valves or mains that cross but do not intersect, or are connected by small-diameter connecting pipes.

K. **Elevations.** Nodal elevations are required by models. Generally, the model results are relatively insensitive to minor errors in elevations, thus this parameter should be one of the last to be modified.

L. **Pipe length.** Pipe lengths are generally scaled from maps, thus there is minimal uncertainty. Another source might be computerized and database mapping systems.

Once an acceptable level of hydraulic calibration of the model has been achieved, parameter relationships can be established. In the AWWARF/USEPA project, for example, the chlorine decay relationship in EPANET based on both bulk decay and pipe-wall decay was used, while a first-order growth relationship was used in modeling TTHMs. In establishing chlorine decay values for the bulk decay Kb and the wall demand Kw, various formal and informal search methods were employed.

BEST-FIT VALUES

The following discussion provides some general insights into the procedures used to determine the best-fit parameters, using chlorine decay as an example. Generally, a good starting point for the bulk chlorine decay rate is derived from controlled bottle decay tests. In these tests, samples of the water are stored in headspace-free, darkened bottles at ambient water temperature.

Individual bottles are then opened at different times, chlorine residuals measured, and the values plotted against time. For a first-order decay, the resulting curve should be a straight line on semilog paper with the slope representing the decay rate. Frequently, however, the curve is concave, convex, or piecewise linear. Therefore, estimation of the best-fit, first-order decay rate might require some judgment and would be handled by determining a range of decay assumptions to be tested in the model.

Methods for guiding the search for the best-fit values of the bulk and wall decay parameters may be classified as follows:

- *Informal adjustment.* The modeler may informally adjust the parameters relying on experience and judgment to lead to an acceptable set of parameters. This method has been most successful in systems dominated by bulk decay and little wall demand.

- *Relationship between pipe characteristics and wall decay parameters.* It is believed that certain pipe characteristics lead to increased wall demand. For example, pipes of a particular age or material may be more susceptible to chlorine demand at the wall or a relationship may exist between the roughness coefficient and wall demand. Using such postulated relationships, the wall demand parameter may be calculated for each pipe and the resulting predicted chlorine residuals compared to field results.

- *Geographically zoned wall decay parameters.* Pipe location may also serve as a surrogate for various factors that affect wall demand. Methods for grouping pipes and assigning and testing alternative wall demand parameters were used in one of the AWWARF/USEPA studies to derive best-fit parameters.

The methods associated with estimating TTHM formation were much more fundamental. Formation rates were assumed to be a function of the source water. Alternative formation rates were informally investigated to determine the best fit.

For both bulk chlorine decay and TTHM formation, significant difficulties arose when there were multiple sources of finished water with significantly varying characteristics. Under these circumstances, for those areas where the waters mixed, the limitations of the EPANET model (and all other commercially available models) to track only a single constituent at a time limited further investigation.

Components of a Sampling Study

Field data collection can be characterized according to the following four types: source water data, system data, tracer data, and systems operation data. Each type of data collection technique will be described.

CHARACTERIZATION OF SOURCE WATERS

A complete characterization of the source water should be performed at each study site. When there are multiple sources, each source is characterized and, if the study area is a significant distance from the source, characterization of samples from one or more locations within the study area should be performed. The constituents analyzed at each site will probably vary slightly due to availability of laboratory facilities and the nature of the water. In the AWWARF/USEPA study, laboratory analysis was generally performed by USEPA, individual utility laboratories, commercial laboratories, and state laboratories. Specific protocols were followed to ensure that the samples were preserved prior to analysis.

SYSTEM DATA

In the North Marin study, free chlorine residual concentrations were taken at all locations and supplemented with total chlorine residual concentrations at some or all sampling locations. Concentrations were measured in the field using various meters using the N, N-diethyl-p-phenylenediamine (DPD) method. In all cases, the meters were precalibrated or checked against chlorine standards. More complex methods for measuring chlorine based on titration were rejected because of the time, cost, and difficulty involving these techniques. Continuous chlorine analyzers were not available for use in the field. Trihalomethane (total and individual species) samples were collected at selected stations and analyzed in the laboratory.

In most cases, temperature readings were taken using digital thermometers. Oxidation–reduction potential (ORP) was measured at selected stations during some of the studies using pocket ORP probes.

TRACER STUDIES

The formation and decay of contaminants in distribution systems is a function of time. Therefore, velocities in a network must be represented accurately. Traditionally, hydraulic calibration includes matching pressures throughout a network, use of C (Hazen–Williams roughness coefficient) values derived from field tests, and matching reservoir tank levels and known flows. Though such methods do improve the reliability of velocity predictions in a network model, they do not ensure that the velocities are adequately modeled. For example, it can be shown that if there is uncertainty in actual pipe diameters because of encrustation or poor record keeping, a model that has been calibrated to faithfully reproduce field pressure measurements will not give accurate velocity information.

Because of the uncertainties surrounding normal hydraulic calibration, traditional hydraulic calibration methods in the AWWARF/USEPA study were supplemented, when feasible, with field tracer methods. With tracer methods, a nontoxic conservative constituent is injected into the distribution system over a fixed period and the movement of the constituent is then monitored throughout the system. For example, in the AWWARF/USEPA study, tracer studies were performed in tandem with collecting chlorine and THM data in order to assist in calibrating the model hydraulically over the same period in which the chlorine and THM data were collected.

The most commonly used conservative tracer is fluoride. In those systems where fluoride is routinely added to the finished water (i.e., SCCRWA), the fluoride feed was turned off for a period of time and the resulting fluoride concentrations were monitored throughout the study area. The fluoride was subsequently turned on and the "front" of fluoridated water was then traced through the system. Generally, fluoride was turned off long enough so that the fluoride front would reach the furthest point of the system being studied.

For those systems in which fluoride is not routinely added, provisions can be made to inject fluoride into the water as it enters the study area. Based on experience, it has been observed that for those systems there may be some loss when fluoride is first added to the pipe wall. In order to reduce this effect, it was found that preconditioning of the system by feeding fluoride for a period of a week prior to the study was effective. Fluoride concentrations were measured using electrode methods in the laboratory.

Alternatives to fluoride include constituents normally injected into the system or natural characteristics from multiple sources. In one case (i.e., North Marin), sodium hydroxide (NaOH) is added at one water source for pH control. By intermittently turning sodium hydroxide off and on, the movement of water from that source and the other source was monitored. In another system (i.e., North Penn), the various water sources had significantly different characteristics (e.g., hardness, THMs, and total to free chlorine ratio). By operating sources at different rates, the water from these systems was traced.

OPERATING DATA

Another type of information required to characterize a water system during a sampling study is system operating data. These data include temporal records for pumping, tank and/or reservoir water level elevations, transfer rates between zones, water usage, and, in some cases, pressures. Information should be collected in either digital or analog form, and, in some cases, can be manually read from meters.

Analytical Methods

A key factor in conducting sampling studies is the selection of chemical and physical analytical methods. The basic laboratory and field values collected and the appropriate methods follow.

CHLORINE RESIDUAL MEASUREMENTS

Free available chlorine residual is the sum of the two chlorine species—hypochlorous acid (HOCl) and hypochlorite ion (OCL^-). Free chlorine is measured as the sum of these two species. Their relative distribution is a function of the temperature and pH with the HOCl (at 20°C) being 100 percent at about pH 4.9, 50 percent at about pH 7.6, and 0 percent (100 percent OCL^-) at about pH 11. This relationship is significant since HOCl has 150–300 times the disinfection power of OCL^-, depending on temperature.

Colorimetric Methods

The most popular and widely used colorimetric method for measuring free (and combined) chlorine residuals is the DPD method. Addition of the DPD reagent to the sample results in the formation of a colored complex that is a function of the chlorine concentration.

Amperometric Methods

Amperometric titration methods were among the earliest to be developed and are still considered the standard for laboratory use, accuracy comparisons, and precision estimates. Analysts must be thoroughly familiar with the procedure to obtain results of acceptable quality.

Electrode Methods

Electrode methods offer the possibility of in-situ continuous monitoring of chlorine residuals useful to utility operators trying to optimize chlorine dosages on a real-time basis to meet regulatory limits in the distribution system. They are invaluable for distribution system modelers collecting data to calibrate distribution network water quality models.

TRIHALOMETHANE MEASUREMENTS

Several methods are available for measuring THMs, including the liquid–liquid-extraction gas chromatographic method and the purge and trap gas chromatographic–mass spectrometric (GC–MS) method.

Liquid–Liquid-Extraction Gas Chromatographic Method

This method is intended only for the determination of the four THMs in finished drinking water or in source water. It is very simple and precise for these compounds. For other compounds, confirmation with GC–MS is normally required. This was the technique most often used for studies described in this book.

Purge-and-Trap, Gas Chromatographic–Mass Spectrometric (GC–MS) Method

This is a generalized method for measuring a wide range of volatile organic compounds (VOCs). It is a sophisticated method that should be used only by highly experienced analysts.

FLUORIDE MEASUREMENTS

Methods available for fluoride measurement include the SPADNS (sodium 2-[parasulfophenylazo]-1,8-dihydroxy-3,6-naphthalene disulfonate) colorimetric method, the fluoride electrode method, the complexone distillation method, and the ion chromatography method. The SPADNS and the electrode method are generally the most satisfactory and will be briefly discussed.

SPADNS Method

The SPADNS method is a colorimetric method based on the reaction between fluoride and a zirconium-dye lake. Fluoride reacts with the dye lake, dissociating a portion of it into a colorless complex anion (ZrF_6^{2-}) and the dye. As the amount of fluoride increases, the color produced becomes progressively lighter.

Fluoride Electrode Method

The fluoride electrode is a laser-type doped lanthanum fluoride crystal that senses the potential established by fluoride solutions of different concentrations. By measuring this potential difference with a pH meter with respect to a standard calomel reference electrode, it is possible to measure fluoride concentration in an analytical range of 0.1 to 10 mg/L. The fluoride electrode method offers the advantage of convenience and speed but must be calibrated frequently for reliable results. A series of samples analyzed by 111 analytical laboratories yielded a relative error ranging from 0.2 to 0.7 percent. The most prevalent interfering ions are aluminum and iron with alkalinity and sulfate of concern in higher concentrations. Buffers such as cyclohexylenediaminetetraacetic acid (CDTA) and fluoroborates can be used to minimize the effect of interfering substances.

References

Clark, R.M., B.W. Lykins, J.C. Block, L.J. Wymer, and D.J. Reasoner. 1994a. Water Quality Changes in a Simulated Distribution System. *Jour. Water Supply Research and Technology–Aqua*, 43(6):263–277.

Clark, R.M., G. Smalley, J.A. Goodrich, R. Tull, L.A. Rossman, J. Vasconcelos, and P. Boulos. 1994b. Managing Water Quality in Distribution Systems: Simulating TTHM and Chlorine Residual Propagation. *Jour. Water Supply Research and Technology–Aqua*, 43(4):182–191.

Vasconcelos, J.J., P.F. Boulos, W.M. Grayman, L. Kiene, O. Wable, P. Biswas, A. Bhari, L.A. Rossman, R.M. Clark, and J.A. Goodrich. 1996. *Characterization and Modeling of Chlorine Decay in Distribution Systems.* Denver, Colo.: American Water Works Association Research Foundation and American Water Works Association.

CHAPTER

8

C H A P T E R

Modeling Waterborne Disease Outbreaks

One of the recommendations from the 1991 conference held by the US Environmental Protection Agency (USEPA) and the American Water Works Association Research Foundation (AWWARF) on water quality modeling was to develop water quality modeling techniques that could be applied to waterborne disease outbreaks. This chapter includes two case studies of modeling applications in waterborne disease outbreaks, one briefly and the other in some detail.

The Cabool, Mo., Outbreak

The first opportunity to attempt to apply water quality models to waterborne outbreak investigations occurred between Dec. 15, 1989, and Jan. 20, 1990, in Cabool, Mo. (population 2,090) (Geldreich 1996, Geldreich et al. 1992). During the outbreak, residents and visitors in Cabool experienced 243 cases of diarrhea (85 bloody) and 6 deaths. The illness and deaths were attributed to the pathogenic agent *Escherichia coli* serotype 0157:H7. At the time of the outbreak, the water source was untreated groundwater. Shortly after the outbreak, USEPA sent a team to conduct a research study to determine the underlying cause of the outbreak.

Exceptionally cold weather prior to the outbreak contributed to two major water system line breaks and 43 water meter replacements throughout the city area. The sewage collection lines in Cabool were generally located away from the drinking water distribution lines, but did cross or were near water lines in several locations. At the time of the outbreak, stormwater drained via open ditches along the sides of the streets and roads. During heavy rainfalls, sewage was observed to overflow manhole covers and flow into streets, parking lots, and residential foundations.

Hydraulic and water quality models were applied to examine the movement of water and contaminants in the system. Steady state scenarios were examined and a dynamic analysis of the movement of water and contaminants associated with meter replacement and the aforementioned breaks was conducted. Typical demand patterns were developed from available meter usage for each service connection and it was found that the water demand was 65 percent of the average well production, indicating inaccurate meters, unmetered uses, and a high water loss in the system.

The modeling effort revealed that the pattern of illness occurrence was consistent with water movement patterns in the distribution system assuming two water line breaks. It was concluded, therefore, that some disturbance in the system, possibly the two line breaks or 43 meter replacements, allowed contamination to enter the water system. Analysis showed that the simulated contaminant movement covered 85 percent of the infected population.

The model applied to the Cabool outbreak was the dynamic water quality model (DWQM) introduced on page 18. With the increasing sophistication of water quality propagation modeling, it has become possible to apply these types of models to waterborne disease outbreaks even more easily then during the Cabool outbreak.

The Gideon, Mo., Outbreak

Another opportunity to model an outbreak occurred in December 1993 in Gideon, Mo., when six to nine cases of diarrhea were reported at a nursing home (Clark et al. 1996). After an initial investigation by the Missouri Department of Health (DOH), the Missouri Department of National Resources (DNR) was contacted and water samples were taken at various points in the system between Dec. 17 and 21, 1993. Several samples were positive and yielded 1–6 total coliforms (TC) per 100 mL and a few samples were fecal coliform (FC) positive. Several other samples yielded results that were too numerous to count (TNTC) for coliforms and were also FC positive.

Original speculation regarding the cause of the outbreak focused on a water tank located on private property. The tank was constructed in 1930 and appeared to be heavily rusted and in an obvious state of disrepair. This tank, connected via a backflow-prevention valve to the city water system, was used primarily for fire protection at the Cotton Compress, a local cotton baling industry.

On Jan. 14, 1994, a USEPA field team, in conjunction with the Centers for Disease Control and Prevention (CDC) and the state of Missouri initiated a field investigation that included a sanitary survey and microbiological analyses of samples collected on site. A system evaluation was conducted in which the EPANET computer model was used to develop various scenarios to explain possible contaminant transport in the Gideon system.

BACKGROUND

The municipal system had two elevated tanks. Tank capacities were 50,000 gal (189,000 L) and 100,000 gal (378,000 L). A third tank (previously described) was located on the Cotton Compress property. It had a volume of 100,000 gal (378,000 L). Both 100,000-gal (378,000-L) tanks had broad flat roofs, while the smaller municipal tank had a much steeper pitch.

The Gideon municipal water system was originally constructed in the mid-1930s and obtained water from two adjacent 1,300-ft (396-m) wells. Water from the wells was not disinfected at the time of the outbreak. The distribution system consisted primarily of small diameter (2-, 4-, and 6-in. [50-, 100-, and 150-mm]) unlined steel and cast-iron pipe. Tuberculation and corrosion were a major problem in the distribution pipes. Raw water temperatures were unusually high for a groundwater supply (58°F [14°C]) because the system overlies a geologically active fault. Under low-flow or static conditions, the water pressure was close to 50 psi (344 kPa). However, under high-flow or flushing conditions the pressure dropped dramatically as will be discussed later. These sharp pressures drops were evidence of major problems in the Gideon distribution system.

In the Cotton Compress yards, water was used for equipment washing, in restrooms, and for consumption. The pressure gradient between the Gideon system and the Cotton Compress system was such that the private storage tank would overflow when the municipal tanks were filling. To prevent this from occurring, a valve was installed in the influent line to the Cotton Compress tank. This same pressure differential kept water in the Cotton Compress tank unless there was a sudden demand in the warehouse area. The entire Cotton Compress water system was isolated from the Gideon system by a backflow-prevention valve. There were no residential water meters in the Gideon system and residents paid a flat service rate ($11.50 per month) for both water and sewage service. The municipal sewage system operated by gravity flow with two lift stations and as of Dec. 31, 1993, served 429 households.

IDENTIFICATION OF THE OUTBREAK

On November 29, the DOH became aware of two high school students from Gideon who were hospitalized with culture-confirmed salmonellosis (Clark et al. 1996b). Within two days, five additional patients living in Gideon were hospitalized with salmonellosis (one student, one child from a day care, two nursing home residents, and one visitor to the nursing home). The state Public Health Laboratories identified the isolates as dulcitol negative *Salmonella* and the CDC laboratories identified the organism as serovar *typhimurium*. Interviews conducted by the DOH suggested that there were no food exposures common to a majority of the patients. All of the ill persons had consumed municipal water.

The Missouri Department of Natural Resources (DNR) was informed that the DOH suspected a water supply link to the outbreak. Water samples collected by the DNR on December 16 were positive for FC. On December 18, the city of Gideon, as required by the DNR, issued a boil water order. Signs were posted at city hall and in the grocery store, and announcements were made over two area radio stations.

Several water samples collected by DNR on December 20 were also found to be FC positive. On December 23, a chlorinator was placed on-line at the city well by DNR, and nine samples were collected by the DOH and DNR from various sites in the distribution system. None of the samples contained chlorine but one sample collected from a fire hydrant was positive for dulcitol-negative *Salmonella* serovar *typhimurium* (Table 8-1). Figure 8-1 shows the location of the sampling points and identifies those points that yielded fecal positive results from the DNR/DOH survey. It is interesting to note from Table 8-1 that most of the sampling points that were coliform positive were also fecal coliform positive. Multiple entries in a column indicate repeat samples.

The Missouri DOH informed the CDC about the outbreak in Gideon in early December and requested information about dulcitol negative *Salmonella* serovar *typhimurium*. On December 17, DOH informed CDC that contaminated municipal water was the suspected cause of the outbreak and on December 22 invited CDC to participate in the investigation. Fliers explaining the boil order, jointly produced by DOH and DNR, were placed in the mailboxes of all of the homes in Gideon on December 29 and the privately owned water tower was physically disconnected from the municipal system on December 30. The DNR mandated that Gideon permanently chlorinate their water system. At the end of the study, USEPA provided input to DNR on the criteria necessary to lift the boil water order.

Through Jan. 8, 1994, DOH had identified 31 laboratory-confirmed cases of salmonellosis associated with the Gideon outbreak. The state Public Health Laboratories identified 21 of these isolates as dulcitol negative *Salmonella* serovar *typhimurium*. Fifteen of the 31 culture-confirmed patients were hospitalized (including two patients hospitalized for other causes who developed diarrhea while in the hospital). The patients were admitted to 10 different hospitals. Two of the patients had positive blood cultures, seven nursing home residents exhibiting diarrheal illness died, four of whom were culture confirmed (the other three were not cultured). All of the culture-confirmed patients were exposed to Gideon municipal water.

Ten culture-confirmed patients did not reside in Gideon but all traveled to Gideon frequently to either attend school (eight), use a day care center (one), or work at the nursing home (one). The earliest onset of symptoms in a culture-confirmed case was on November 17 (this patient was last exposed to Gideon water on November 16). A CDC survey indicated that approximately 44 percent of the 1,104 residents, or almost 600 people, were affected with diarrhea between Nov. 11 and Dec. 27, 1993. Nonresidents who drank Gideon water during the outbreak period experienced an attack rate of 28 percent (Angulo et al. 1997).

POSSIBLE CAUSES

The investigation clearly implicated consumption of Gideon municipal water as the source of the outbreak. Speculation focused on a sequential flushing program conducted on November 10 involving all 50 hydrants in the system. The program was started in the morning and continued through the entire day. Each hydrant was flushed for 15 minutes at an approximate rate of 750 gpm (2,838 L/min). It was observed that the pump at well 5 was operating at full capacity during the flushing program (approximately 12 hours), which would indicate that the municipal tanks were discharging during this period. The flushing program was conducted in response to taste-and-odor complaints.

It was hypothesized that taste-and-odor problems may have resulted from a thermal inversion that may have taken place due to a sharp temperature drop prior to the day of the complaint. If stagnant or contaminated water was floating on the top of the tank, a thermal inversion could have caused this water to be mixed throughout the tank and to be discharged into the system resulting in taste-and-odor complaints (Mendis et al. 1976). As a consequence, the utility initiated a city-wide flushing program. Turbulence in the tank from the flushing program probably stirred up the tank sediments, which were transported into the distribution system. It is likely that the bulk water and/or the sediments were contaminated with *Salmonella* serovar *typhimurium*.

During the USEPA field visit, a large number of pigeons were observed roosting on the roof of the 100,000-gal (378,000-L) municipal tank. Shortly after the outbreak, a tank inspector found holes at the top of the Cotton Compress tank, rust on the tank, and rust, sediment, and bird feathers floating in the water. According to the inspector, the water in the tank looked black and was so turbid he could not see the bottom. Another inspection, conducted after USEPA's field study, confirmed the disrepair of the Cotton Compress tank and also found the 100,000-gal (378,000-L) municipal tank in such a state of disrepair that bird droppings could, in the opinion of the inspector, have entered the stored water. Bird feathers were in the vicinity or in the tank openings of both the Cotton Compress and the municipal tank.

It was initially speculated that the backflow valve between the Cotton Compress and the municipal system might have failed during the flushing program. After the outbreak, the valve was excavated and found to be working properly. The private tank was drained accidently during an inspection after the outbreak so it was impossible to sample water in the tank bowel. However, sediment in the private tank contained *Salmonella* serovar *typhimurium* dulcitol negative organisms as did samples from a hydrant and culture-confirmed patients. The *Salmonella* found in the street hydrant (304 6th St.) matched the serovar of the patient isolate when analyzed by the CDC laboratory comparing DNA fragments using pulse field gel electrophoresis (PFGE). (PFGE is a molecular biology procedure used to help identify strains of bacteria [normally the same species]. It has gained wide acceptance in epidemiological studies to trace the origin of organisms.) The isolate from the tank sediment, however, did not provide an exact match with the other two isolates.

MODELING WATERBORNE DISEASE OUTBREAKS

Table 8-1 Coliform analysis of samples collected by **MDNR** and **MDOH** between Nov. 15 and Dec. 23, 1993

Sample No.	Sample Location	11/15/93 CFU*/100 mL	11/22/93 CFU/100 mL	12/6/93 CFU/100 mL	2/13/93 CFU/100 mL	2/16/93 CFU/100 mL	12/17/93 CFU/100 mL	12/20/93 CFU/100 mL	12/21/93 CFU/100 mL	12/22/93 CFU/100 mL	12/23/93 CFU/100 mL
						DATE					
Source Water											
15	Well 5	—	—	—	—	—	—	—	<1/<1	—	—
7	Well 1	—	—	—	—	—	<1	—	—	—	—
Distribution System Hydrants											
17	6th St.†										6‡
Service Taps: Residences and Public Buildings											
1	S. Walker§	<1	<1	—	—	—	—	—	—	—	—
2	Gideon Tire§	—	—	<1	—	—	—	—	—	—	—
3	Gideon Airport§	—	—	—	<1	—	—	—	—	—	—
4	105 Anderson Ave.	—	—	—	—	<1	—	—	—	—	—
5	High School	—	—	—	—	—	—	3‡/3‡	6/7	—	—
6	302 N. Railroad	—	—	—	—	—	—	<1/<1	<1	<1	—
8	402 S. Walker	—	—	—	—	—	<1	—	—	—	—
9	Nursing Home	—	—	—	—	—	<1	—	—	—	—
10	209 Gideon St.	—	—	—	—	—	<1	—	—	—	—

table continues next page

Table 8-1 Coliform analysis of samples collected by MDNR and MDOH between Nov. 15 and Dec. 23, 1993 (Continued)

Sample No.	Sample Location	11/15/93 CFU*/100 mL	11/22/93 CFU/100 mL	12/6/93 CFU/100 mL	2/13/93 CFU/100 mL	2/16/93 CFU/100 mL	12/17/93 CFU/100 mL	12/20/93 CFU/100 mL	12/21/93 CFU/100 mL	12/22/93 CFU/100 mL	12/23/93 CFU/100 mL
									DATE		
11	124 Whiterow	—	—	—	—	—	TNTC†/TNTC‡	8‡/6/15	<1/<1	—	—
12	205 Lunback Ave.	—	—	—	—	<1		—	—	—	—
13	Grade School	—	—	—	—	—	6‡	—	—	—	—
14	Compress Interconnection	—	—	—	—	—	—	4/6	—	—	—
16	112 Whiterow	—	—	—	—	—	—	18‡/10‡	<1	—	—
18	401 N. Anderson	—	—	—	—	—	—	—	—	—	—
19	209 Whiterow	—	—	—	—	—	—	—	<1	—	—
20	309 Elkens St.	—	—	—	—	—	—	—	<1	—	—

* Colony-forming units.

† *Salmonella* isolated.

‡ Fecal coliform positives.

§ Denotes regular surveillance sample.

TNTC = too numerous to count.

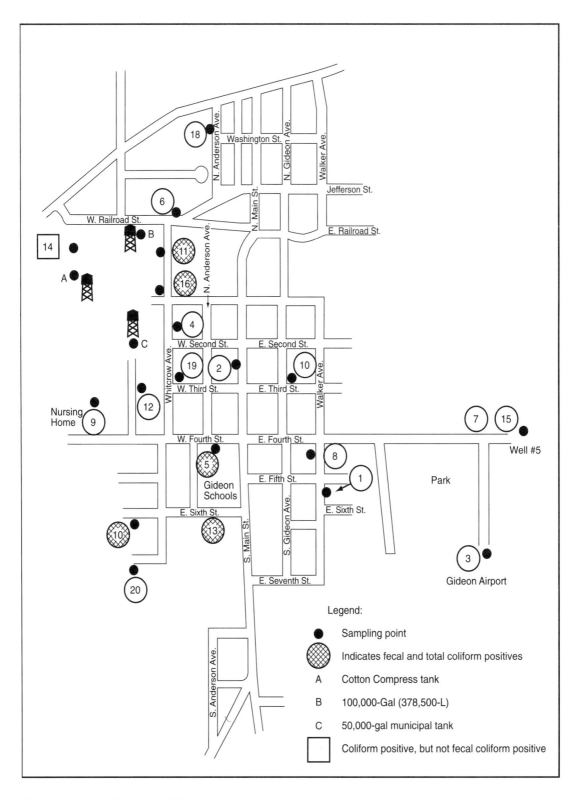

Figure 8-1   DNR and DOH sampling results

USEPA FIELD STUDY

During the course of a cross-connection survey, several locations were identified that were potential cross connections and it was observed that the wellhead areas were not properly protected. An extensive microbiological survey was conducted and the system was found to have high levels of heterotrophic plate count bacteria (HPCs) in various locations.

Sanitary Survey

Although the investigation did not constitute a complete sanitary survey, it did suggest obvious areas of possible concern, some of which are described below. Figure 8-1 shows the sampling sites used for USEPA's microbiological survey.

Wellhead Protection. It was found that agricultural areas surrounding the wells drained across the wellheads and subjected the area to large amounts of runoff. The two wells that supplied source water (designated 5 and 6) were deep (1,300 ft [396.2 m]) artesian and thought to be protected by the underlying geological formations. This protection could not be checked during the field study; it was suggested that the integrity of the aquifer should be investigated.

The two wells operated alternately on a monthly basis. When one well was pumping, the other was turned off. There did not appear to be any backflow devices (air gap breaks) to prevent a siphon from being placed on the unused well pump when the other pump was turned on. This was of concern because the water meter and valve for well 6 was located in a pit that reportedly flooded routinely. The pit filled with runoff from the surrounding fields and had to be pumped out before the meter could be read. Water in the pit could be pulled into the distribution system and the contaminated water distributed throughout the system. When well 6 was pumping, a venturi effect could also pull some of the pit water into the system.

There were some inconsistencies with regard to the free chlorine residual found at several locations. At the nursing home, the residual was 0.2 mg/L, while only a block away on Whiterow, the measured amount was 1.08 mg/L. This indicated a closed pipe or hydraulic obstructions in the system.

Microbiological Survey

In an effort to characterize the microbial quality and to evaluate the effectiveness of applied disinfection as a remedial measure, samples were taken on Jan. 5 and Jan. 6, 1994, from selected sites in the system. Sampling locations included the source water (two artesian wells); fire hydrants near the water tanks, areas of static water, and areas near homes where illness occurred; tap water from a residential service line, and taps in the restrooms of the nursing home and elementary school (Figure 8-1). In addition, surface drainage around well meter boxes, water from a 30-ft (9-m) private well, drainage water

Table 8-2 Summary of sample analyses

Number of Samples	Dates Collected	Dates Received and Analyzed
10	Jan. 5, 1994	Jan. 6, 1994
8	Jan. 6, 1994	Jan. 7, 1994
1	Jan. 12, 1994	Jan. 13, 1994

and sediment from the private tank, and one historical sample collected during the outbreak period were examined. The samples were shipped on ice via overnight carrier to USEPA's laboratory and were analyzed on the day of receipt. Table 8-2 summarizes the sampling history. All samples were analyzed for total coliform bacteria, fecal coliform bacteria, heterotrophic plate count bacteria, and *Salmonella*.

Samples were analyzed for total coliform bacteria by the membrane filter technique using m-Endo LES agar and by the presence–absence procedure using presence–absence (PA) broth (APHA, AWWA, and WEF 1992). Fecal coliform levels were determined by the elevated temperature method using mFC agar in the membrane filter procedure (APHA, AWWA, and WEF 1992). The spread plate technique using R2A agar incubated at 25°C for 7 days was used to enumerate heterotrophic plate count bacteria. All turbid enrichment cultures exhibiting an orange to red color were streaked onto bismuth sulfite agar, brilliant green agar, and xylose lysine deoxycholate agar (APHA, AWWA, and WEF 1992). Target colonies were characterized biochemically and by serological examination. Biochemical characterization of coliform and *Salmonella* isolates was performed using the API 20E multitest kit[*] (APHA, AWWA, and WEF 1992). Two-litre samples for *Salmonella* analysis were concentrated by the membrane filter or diatomaceous earth procedure. Selective enrichment of the concentrated samples was conducted in selenite cystine broth incubated at 41.5°C (APHA, AWWA, and WEF 1992; Spino 1966). Serological identification was accomplished using *Salmonella* polyvalent somatic (0) and polyvalent flagellar (H) antisera[†] (FDA and AOAC International 1992). Appropriate positive and negative controls were included for all analyses.

Coliform Sampling Protocol

Source Water.   Microbiological samples collected on Jan. 6, 1994, from the wells contained no detectable coliforms per 100 mL and less than 200 heterotrophic bacteria per millilitre (Table 8-3). A sample of sump water in the meter well was examined bacteriologically to characterize the extent of this contamination threat. As anticipated, the sample contained a very high density

---

[*] Available from Merieux Vitek, Inc., Hazelwood, Mo.

[†] Available from Difco Laboratories, Detroit, Mich.

of heterotrophic bacteria, estimated to be greater than 100,000 organisms per millilitre. This excessive bacterial population interfered with establishing the density of coliform bacteria present in the sample. A PA coliform test yielded positive results and indicated that coliform bacteria were present.

Tank Storage.   Water quality in the tanks was investigated by sampling at the nearest fire hydrant during a drawdown period from both the 100,000-gal (378,000-L) and 50,000-gal (189,000-L) municipal tanks (samples 14 and 15 in Table 8-3). *Enterobacter cloacae* and a strain of *Aeromonas hydrophila* were not detected at the nearest fire hydrant (sample 15) to the 100,000-gal (378,000-L) tank by standard coliform procedures but were detected in the *Salmonella*-selective differential medium incubated at an elevated temperature (41°C). These results suggest heterotrophic bacteria could have interfered with coliform detection. The nearest hydrant to the 50,000-gal (189,000-L) storage tank contained a significant amount of particulates that limited the membrane filter sample to 20 mL. Three isolates of *Klebsiella oxytoca* were recovered in the sample. The presence of particulate matter suggested that either this hydrant was not adequately flushed or that the tank water supply was static. This finding was consistent with the concept of stagnant water in the tanks.

Pathogen Detection.   All samples were also examined for *Salmonella* because this pathogen was the suspected agent in the Gideon outbreak. *Salmonella* had been found in the stool of one patient and, on Dec. 16, 1993, in water from the fire hydrant on 6th St. (304 6th St., sample 2 in Figure 8-1). No *Salmonella* was detected in any of the samples collected on January 5 and 6 in the public water system, most likely because disinfection was initiated on Dec. 23, 1993.

    Salmonella was found in the sediment collected on January 5 from the riser pipe of the Cotton Compress water storage tank. As mentioned previously, the water tank had been physically disconnected from the public water supply on Dec. 30, 1993, and accidentally drained in the process. Residual water in the riser pipe did not contain detectable *Salmonella*, but both residual water and sediment samples did contain the coliform *Enterobacter cloacae*. Sources of *Salmonella* and coliforms may have been from feces of pigeons observed roosting in the tower vents. Among bird populations, there are always a few individuals shedding these organisms (Fennel, James, and Morris 1974; Koplan et al. 1978; Jones, Smith, and Watson 1978; and Mutter 1990).

System Evaluation

The purpose of the system evaluation was to study the effects of distribution system design and operations, demand, and hydraulic characteristics on the possible propagation of contaminants in the system (Clark et al. 1996). Given the evidence from the survey and the results from the valve inspection at the

Cotton Compress, it was concluded that the most likely contamination source was bird droppings in the large municipal tank. Therefore, the analysis concentrated on propagation of water from the large municipal tank in conjunction with the flushing program. This did not rule out other possible sources of contamination, such as cross connections.

Table 8-3 Microbiological characteristics of the Gideon, Mo., public water supply

Sample No.	Sample Location	Date	Free Cl$_2$, mg/L	HPC, per mL	Total Coliform, per 100 mL MF P/A		Fecal Coliform, per 100 mL	Species
Source Waters								
12	Well 5	Jan. 5	0.00	<10	<1	–	<1	
13	Well 6	Jan. 5	N.D.*	140	<1	–	<1	
Municipal Water Tanks								
15	100,000 gal	Jan. 5	0.11	560	<1	–	<1	
14	50,000 gal	Jan. 5	0.11	7,000	20†	–	<5†	Kleb. oxytoca
Distribution System Hydrants								
1	So. Anderson	Jan. 5	0.04	45	<1	–	<1	
2	304 6th St.	Jan. 5	0.02	20	<1	–	<1	
3	120 Haven Ave.	Jan. 5	0.02	25,000	1	–	<1	C. freundii
4	Jefferson St. Street hydrant	Jan. 5	0.11	24,000‡	10	–	<1	C. freundii E. cloacae
5	2nd St.	Jan. 5	0.04	24,000‡	<1	+	<1	C. freundii
Service Taps—Residence, Public Buildings								
22	122 Whiterow St.	Jan. 6	1.08	5	<1	–		
20	Nursing Home	Jan. 6	1.12	20	<1	–	<1	
21	Grade School	Jan. 6	0.21	<10	<1	–	<1	
Compressor Water Tank								
10	Drainage	Jan. 5	0.00	9,000	<5†	–	<5†	E. cloacae
8	Sediment§	Jan. 5	0.00	320	<100†	+	<100†	E. cloacae
6	Fire hydrant	Jan. 5	0.15	95	<1	–	<1	
Surface Water								
23	Payne well	Jan. 6	N.D.	6,600	1	+	<1	E. agglomerans E. intermedium
24	Meter box sump	Jan. 12	N.D.	>100,000	<1	+	<1	C. freundii

* Not done.

† Density adjusted because sample volume was limited (1, 10, or 20 mL) due to heavy particulates.

‡ *Chromobacterium* detected in HPC—soil organism.

§ *Salmonella* serovar *typhimurium* isolated.

The system's layout, demand information, pump characteristic curves, tank geometry, flushing program, and other information needed for the modeling effort was obtained from maps, demographic information, and numerous discussions with consulting engineers and city and DNR officials. EPANET was used to conduct the contaminant propagation study (Rossman 1994).

NETWORK LAYOUT

Overlays were produced from the distribution system map (developed in May 1981) obtained from the city of Gideon, and links and nodes were identified according to the EPANET format. Links were first identified using a city map. Nodes were then identified based on pipe intersections, major changes in alignment, or dead ends. Hydrants were identified as separate nodes in order to provide the basis for replicating the flushing program. Houses and businesses were aggregated to provide system demand. A few nodes were inserted along uninterrupted links to accommodate hydraulic demands from the houses. The three tanks and the well in operation in November were identified separately, according to EPANET requirements.

NODE CHARACTERISTICS

Elevations for the nodes were derived from the 1978 US Geological Survey (USGS) Madrid topographic map. Hydraulic demands for each node were calculated based on the number of houses assigned to each node. Household usage was calculated based on CDC reports of 1,104 people living in 429 homes (2.6 people/house) and an assumption of 75 gal (280 L) of water used per person per day. The daily water use patterns (drinking, bathing, washing, cooking, and lawn watering) developed in an earlier study in Cabool, Mo., were used for the residences and the schools (Geldreich et al. 1992). Daily demands for the nursing home were based on the CDC report of 68 residents and 62 staff, assuming the same consumption rate. A lower rate of 20 gal (76 L) used per student per day was estimated for a node serving the school area.

The calculated consumption rates were very close to those used in Cabool, Mo., where meter and water use information was available for each residence, the nursing home, and schools. Pump records indicated an average of 130,000 gal (492,000 L) of water pumped daily during the outbreak in November. Daily consumption estimated using the per capita assumptions were approximately 77 percent of the daily pumpage. This difference was not unexpected nor excessive given experience in other communities. Water leaks and/ or unaccounted-for users could have accounted for this discrepancy as well as low per capita estimates. Table 8-4 summarizes some of the key values used as part of the EPANET program.

TANK AND PUMP CHARACTERISTICS

Based on the EPANET format, the tanks were identified by number. Tank 200 (T200) is the 50,000-gal (189,000-L) municipal tank, tank 300 (T300) is the 100,000-gal (378,000-L) municipal tank, and tank 400 (T400) is the 100,000-gal (378,000-L) Cotton Compress tank (Figure 8-2).

It was learned that when pressure from a gauge near T200 dropped below 60 psi (413 kPa), the pump at the well was automatically turned on. Several pressure studies conducted at selected hydrants in the system were used to help calibrate EPANET. A pump head curve was provided by DNR. Table 8-5 contains the pressure drop readings obtained from this study by attaching a pressure gauge to an outside faucet near the hydrant to be tested. The hydrants were flushed for 15 minutes at 750 gpm (2,838 L/min) and the pressure recorded.

Table 8-4   Assumptions used in analysis

Item	Value
Number of homes in Gideon	429
Number of residents in Gideon	1,104
Persons/households in Gideon	2.6
Average daily consumption[*]	130,000 gal (492,000 L)
50,000-gal (189,000-L) tank (T200)	
Height	29 ft (8.8 m)
Diameter	18 ft (5.5 m)
100,000-gal (378,000-L) tank (T300)	
Height	24.5 ft (7.5 m)
Diameter	30 ft (9.1 m)
Cotton Compress	
100,000-gal (378,000-L) tank (T400)	
Height	33 ft (10.1 m)
Diameter	24 ft (7.3 m)

* Includes drinking, bathing, washing, cooking, lawn watering, etc.

Table 8-5   Pressure test results

| Hydrant Number | Pressure | | | |
	Static		Dynamic	
	psi	kPa	psi	kPa
4	58	400	7	48
9	53	365	8	55
49	50	345	18	124

LINK CHARACTERISTICS

Links were calculated from the 1 in.:200 ft (25 mm:61 mm) scale map. Diameters were derived from the 1951 system map. Roughness coefficients were estimated based on the general age and location of the pipes and from the pressure readings shown above.

SYSTEM PERFORMANCE

EPANET was calibrated by simulating flushing at the hydrants shown in Table 8-5, assuming a discharge of 750 gpm (2,850 L/min) for 15 minutes. The C factors were adjusted until the head loss in the model matched head losses observed in the field.

The hydraulic scenario was initiated by running the model for 48 hours. The water level reached 400.59 ft (122 m) in T400 (Cotton Compress tank), 400.63 ft (122 m) in T300, and 400.66 ft (122 m) in T200.

At 8:00 a.m. on the third day, the simulated flushing program was initiated by sequentially imposing a 750-gpm (2,850-L/min) demand on each hydrant, 1 through 50, for 15 minutes. The entire process took 12.5 hours. Using the TRACE option in EPANET, the percentages of water from both municipal tanks were calculated at each node over a period of 72 hours. Based on the findings from excavating the backflow-prevention valve, the impact of flows from T400 (Cotton Compress) were not considered in the simulation.

It was found from the simulation that the pump operated at over 800 gpm (3,028 L/min) during the flushing program and then reverted to cyclic operation thereafter. The tank elevations for both municipal tanks fluctuate, and both the tanks discharged during the flushing program. At the end of the flushing period, nearly 25 percent of the water from T300 had passed through T200, where it was again discharged into the system.

Pressure drops during the flushing program were simulated at the hydrants used for calibration. The model predicted dramatic pressure drops during the flushing program and nodes 4 and 49 showed negative pressures, which were considered as zero. It was believed that these results replicated the conditions that existed during the flushing program closely enough to provide the basis for an analysis of water movement in the system.

WATER MOVEMENT AND OUTBREAK PATTERN

Perhaps the key to the Gideon outbreak pattern and a major clue as to the most likely source of contamination is given by tracing the movement of water from both municipal tanks during the flushing program. Figure 8-2 shows the hourly movement of water from T300 in the system during the first four hours of the flushing period.

Water movement from T200 tends to dominate the area immediately around the tank during the first few hours of the flushing period until it is drawn down. Water from T300 initially supplies most of the northern and

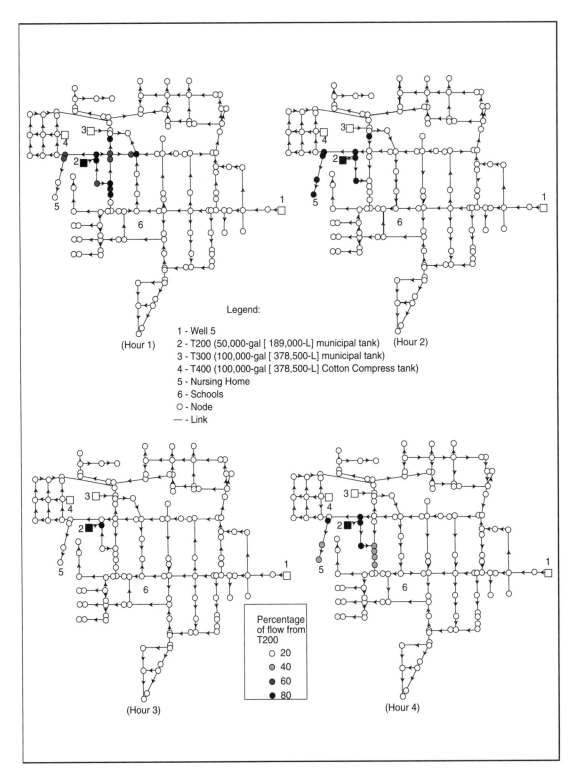

Figure 8-2 Water movement from T200 during hours 1–4 of 72-hour simulation period (T400 valve closed)

western portions of the system. Almost all of the water in the southern and eastern portions of the system is supplied by the well.

Water movement was determined at 6-hour intervals for both T200 and T300, respectively, starting at a point 24 hours into the simulation period or when normal operation has resumed. Water from both T200 and T300 reaches virtually all of the system under normal operation, with approximately 25 percent of the water in T200 passing through the presumably contaminated T300.

The percentage of water from T300 was determined at selected nodes in the system. During the flushing program there were periods when 100 percent of the water at nodes near the school, the hydrant yielding the positive *Salmonella* serovar *typhimurium* (sample 2 in Table 8-1), and the nursing home were from T300. At all nodes, water from T300 is present at some time during the three-day simulation period.

CONTAMINANT PROPAGATION

Data from the simulation study, the microbiological surveillance data in Tables 8-1 and 8-2, and the outbreak data can be used to provide insight into the nature of both general contamination problems in the system and into the outbreak itself. Table 8-1, Figure 8-1, and the water movement patterns show that the majority of the special samples that were coliform and fecal coliform positive occurred at points that lie within the zone of influence of T200 and T300. During both the flushing program and for large parts of normal operation, these areas are predominately served by tank water. This might indicate that the tanks are the source of the fecal contamination because there were positive FC samples prior to chlorination.

Data from the early cases, in combination with the water movement data, was used to infer the source of the outbreak. Using data supplied by CDC and the water movement simulations, an overlay of the areas served by T200 and T300 during the first six hours of the flushing period and the earliest recorded cases was created (shown in Figure 8-3). As can be seen, the earliest recorded cases and the positive *Salmonella* hydrant sample were found in the area primarily served by T300, but outside the T200 area of influence, during the flushing period. One can conclude that during the first six hours of the flushing period the water that reached the residence and the Gideon School was almost totally from T300. Therefore, it is logical to conclude that these locations should experience the first signs of the outbreak, which makes a strong circumstantial case for T300 as the contamination source.

Figure 8-4 displays the increase in the number of absentees from the Gideon schools during the outbreak period. As can be seen, there was a sudden rise in absentees on November 12, two days after the hydrant flushing program. Figures 8-5 and 8-6 show the progress of disease during the first few days of the outbreak. As mentioned, the disease progressed in an apparently random manner after the first occurrences in the center of the city.

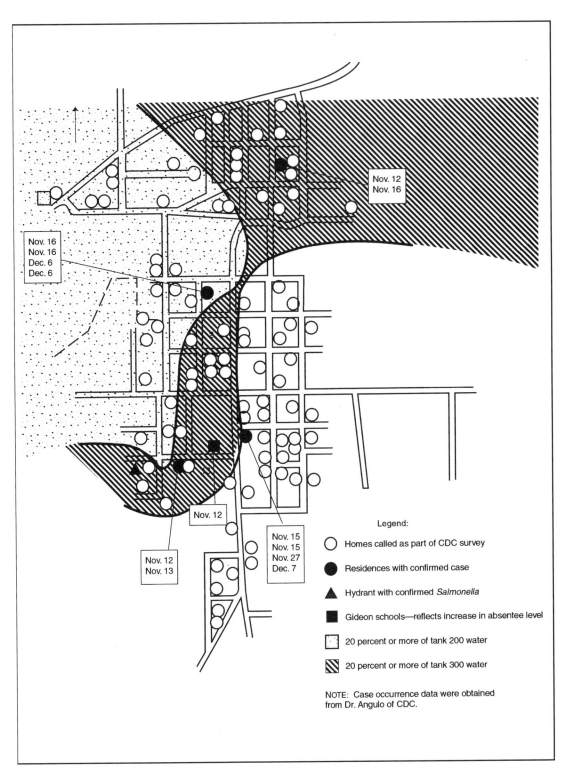

Figure 8-3 Comparison of early confirmed cases and *Salmonella*-positive sample versus penetration of tank water during first six hours of flushing program

These arguments support the hypothesis that the sudden drop in temperature on the night of Nov. 9, 1993, caused a turnover of water in the tank, thereby mixing the contaminated portion of the tank water with the relatively clean portion of the water column. This probably caused taste-and-odor complaints that resulted in the rigorous flushing program on November 10. Taste-and-odor problems had been reported previously but only a limited flushing program in the area of complaints was attempted. The sudden discharge of the water column probably stirred up contaminated sediments in the bottom of this tank and resulted in the outbreak.

Given the history of coliform violations and taste-and-odor complaints, it is reasonable to question whether or not this outbreak was a one-time event or whether the contamination had been occurring routinely or at least periodically. These results may also explain some of the previous coliform violations found in the utility records.

Based on the results of the DNR/DOH sampling program, it is likely that the contamination had been occurring over a period of time, which is consistent with the possibility of bird contamination. If the cause was a single event, the contaminant would most likely have been "pulled" through the system during the flushing program.

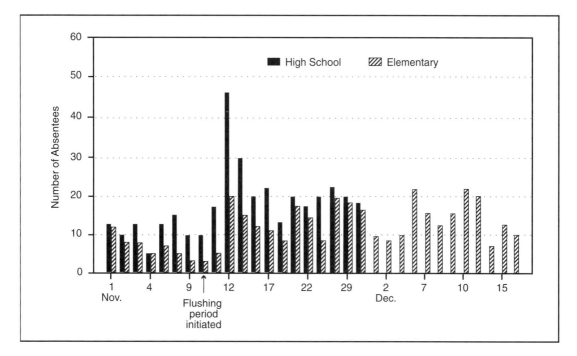

Figure 8-4   Number of absentees in Gideon schools

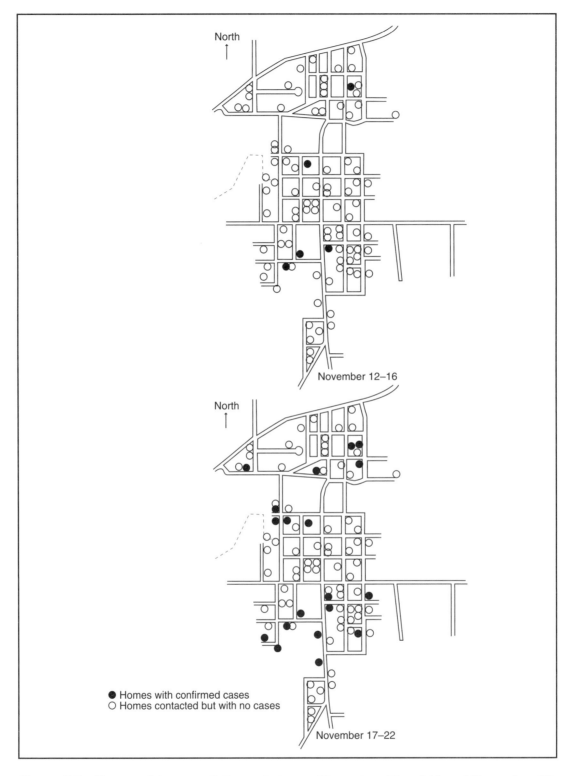

Figure 8-5   Homes with cases of disease between November 12 and 16 and November 17 and 22

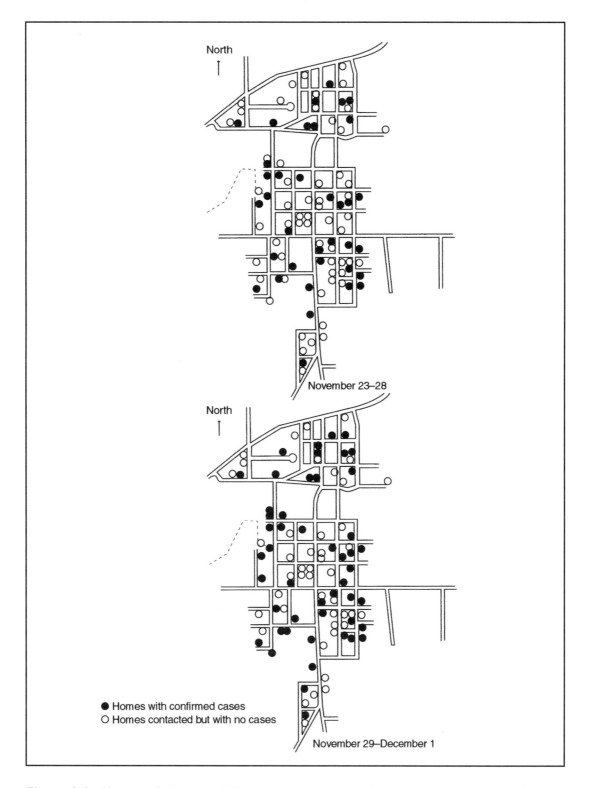

Figure 8-6   Homes with cases of disease between November 23 and 28 and November 29 and December 10

Summary and Conclusions

The density of a pathogenic agent in a contaminated water supply is subject to much variation because of contaminate input, dilution effect of water, and any available disinfectant residual. In the Gideon case study, there was no applied disinfectant and the density of *Salmonella* in the stratified water supply in the tanks was unknown. However, the persistence of this pathogenic agent was extended by the static nature of the stratified water.

Laboratory study of the persistence of this *Salmonella* strain in the Gideon water supply demonstrated that the pathogenic agent was only reduced in density by 30 percent during a four-day period at 15°C. Thus, with repeated new inputs of *S. typhimurium* from infected pigeons, there could be a continuing high level of this pathogen present.

Obviously, this situation provided a sufficient cell density to be an infective dose to many people in the community once the tank water was destratified by an abrupt drop in air temperature and released to the pipe network through the flushing activity near the tanks. Infective dose levels for different *Salmonella* species may vary from 10^1 (*S. typhi*) to more than 10^5 cells (*S. typhimurium*). In addition, there are a number of intervening factors that explain why not all people exposed to a contaminated water source will become ill. Part of this variation in human response relates to water intake per day and the nature of an individual body's defense against pathogen colonization. Individuals at greatest risk are infants, senior citizens, people taking excessive stomach antiacid medicines, alcoholics, persons exposed to radiation or chemical therapy, and those with acquired immune deficiency syndrome (AIDS victims).

It is reasonable to assume, given the pattern and rapidity of the spread of *Salmonella* serovar *typhimurium* in Gideon in November and December of 1993, that a waterborne disease outbreak had occurred. The cause of the outbreak is not obvious but was likely associated with bird contamination in the storage tanks. Given that the only valve connecting the Cotton Compress and the Gideon distribution system was found to be closed, the outbreak is most likely associated with the largest municipal tank, which provided an excellent roosting place for pigeons. A tank inspector observed that the tank was covered with bird feathers, dirt, and droppings. It was also observed that the vents in the tank were designed in such a way as to allow for the possibility of contamination. Finally, it was found that the strain of *Salmonella* serovar *typhimurium* isolated from a patient survived for several days. This finding supports the hypothesis that the source of organisms was in the public water supply tanks and pulled through the system during a vigorous flushing program that started near the tanks.

Water quality modeling played a major role in finding a reasonable hypothesis for the cause of the outbreak.

References

Angulo, F.J., S. Tippen, D.J. Sharp, B.J. Payne, C. Collier, J.E. Hill, T.J. Barrett, R.M. Clark, E.E. Geldreich, H.D. Donnell, and D.L. Swerdlow. 1997. A Community Waterborne Outbreak of Salmonellosis and the Effectiveness of a Boil Water Order. *American Journal of Public Health*, 87(4):580–584.

APHA, AWWA, and WEF. 1992. *Standard Methods for the Examination of Water and Wastewater*, 18th ed. Washington, D.C.: American Public Health Association.

Bacteriological Analytical Manual, 7th ed. Arlington, Va.: Food and Drug Administration and Association of Official Analytical Chemists.

Clark, R.M., E.E. Geldreich, K.R. Fox, E.W. Rice, C.W. Johnson, J.A. Goodrich, J.A. Barnick, F. Abdesaken. 1996. Tracking a *Salmonella* Serovar *Typhimurium* Outbreak in Gideon, Missouri: Role of Contaminant Propagation Modeling. *J. Water Supply Research and Technology-Agua*, 45(4):171–183.

Clark, R.M., E.E. Geldreich, K.R. Fox, E.W. Rice, C.W. Johnson, J.A. Goodrich, J.A. Barnick, F. Abdesaken, J.E. Hill, F.J. Angulo. 1996. A Waterborne *Salmonella Typhimurium* Outbreak in Gideon, Missouri: Results From A Field Investigation. *Intl J Environ Health Research*, 6(3): 187–193.

Fennel, H., D.B. James, and J. Morris. 1974. Pollution of a Storage Reservoir by Roosting Gulls. *Jour. Soc. Water Treat. Exam.*, 23:5–24.

Geldreich, E.E. 1996. *Microbial Quality of Water Supply In Distribution Systems*. Boca Raton, Fla.: Lewis Publishers. 377–385.

Geldreich, E.E., K.R. Fox, J.A. Goodrich, E.W. Rice, R.M. Clark, and D.L. Swerdlow. 1992. Searching for a Water Supply Connection in the Cabool, Missouri Disease Outbreak of *Escherichia coli* 0157:H7. *Water Research,* 84:49–55.

Jones, F., P. Smith, and D.C. Watson. 1978. Pollution of a Water Supply Catchment by Breeding Gulls and the Potential Environmental Health Implications. *Jour. Institution Water Engrs. Sci.*, 32:469–482.

Koplan, J.P., R.D. Deen, W.H. Swanston, and B. Tota. 1978. Contaminated Roof-Collected Rainwater as a Possible Cause of an Outbreak of Salmonellosis. *J. Hyg.*, 81:303–309.

Mendis, N.M.P., P.U. De LaMotte, P.D.P. Gunatillaka, and W. Nagaratnam. 1976. Protracted Infection with *Salmonella bareilly* in a Maternity Hospital. *J. Tropical Med. and Hyg.*, 79:142–150.

Mutler, D.F. 1990. The Pharmacokinetics of Dihydrostreptomysin Sulfate in Domestic Pigeons. *Tierarztl Praf.*, 18:377–381.

Rossman, L.A. 1994. *EPANET Users Manual*. Cincinnati, Ohio: USEPA, Drinking Water Research Division.

Spino, D.F. 1966. Elevated Temperature Technique for the Isolation of *Salmonella* from Streams. *Appl. Microbiol.*, 14:591–596.

Modeling the Effects of Tanks and Storage

A frequently overlooked aspect of water quality and contaminant propagation in drinking water distribution systems is system storage. Although direct pumping could maximize water quality by shortening the transport time between source and consumer, it is rarely used today in systems in the United States (AWWA 1989). Most utilities use some type of ground or elevated system storage to process water at times when treatment facilities would otherwise be idle. It is then possible to distribute and store water at one or more locations in the service area closest to the user.

The principal advantages of distribution storage are that it equalizes demands on supply sources, production works, and transmission and distribution mains. As a result, the sizes or capacities of these elements may be minimized. Additionally, system flows and pressures are improved and stabilized to better serve the customers throughout the service area. Finally, reserve supplies are provided in the distribution system for emergencies, such as fire fighting and power outages.

In most municipal water systems, less than 25 percent of the volume of the storage in tanks is actively used under routine conditions. As the water level drops, the tank controls call for high-service pumps to start in order to satisfy demand and refill the tanks. The remaining water in the tanks (70 to 75 percent) is normally held in reserve as dedicated fire storage.

Storage tanks and reservoirs are the most visible components of a water distribution system but are often the least understood in terms of their effect on water quality. Although these facilities can play a major role in providing hydraulic reliability for fire fighting needs and in providing reliable service, they may also serve as vessels for complex chemical and biological changes that may result in the deterioration of water quality.

This chapter will examine the issue of residence time on water quality and the use of "compartment" models for describing mixing regimes in tanks.

Previous Research

P.V. Danckwerts, one of the first investigators to discuss the concept of distribution functions for residence times, explained how this concept can be defined and measured in actual systems (Dankwerts 1958). When a fluid flows through a vessel at a constant rate, either "plug flow" (no mixing) or perfect mixing is usually assumed. In practice many systems do not achieve either assumption, thus calculations based on these assumptions may be inaccurate. Danckwerts illustrated the use of distribution functions by showing how they can be used to calculate the efficiencies of reactors and blenders and how models may be used to predict the distribution of residence times in large systems.

A.E. Germeles developed a model based on the concept of forced plumes and mixing of liquids in tanks (Germeles 1975). He considered the mixing between two miscible liquids of slightly different density (<10 percent) when one of them is injected into a tank partially filled with the other. A mathematical model for the mixing of the two liquids was developed, from which one can compute the tank stratification. The model was also verified experimentally.

EMPIRICAL STUDIES

Several investigators have conducted field studies and attempted to apply relatively simple models to distribution storage tanks. Kennedy, Moegling, and Suravallop (1993) attempted to assess the effects of storage tank design and operation on mixing regimes and effluent water quality. The influent and effluent flows of three tanks with diameter-to-height ratios ranging from 3.5:1 to 0.4:1 were monitored for chlorine residual. Chlorine levels were also measured within the water columns of each tank. Although chlorine profiles revealed some stratification in tanks with large height-to-diameter ratios, completely mixed models were more accurate than plug–flow models in representing the mixing behavior of all three tanks. These investigators further indicated that the quality of the effluent from completely mixed tanks deteriorated with decreasing volumetric change. The authors found that standpipes were the least desirable tank design with respect to effluent water quality.

Studies conducted by Grayman and Clark (1993) indicated that water quality degrades as a result of long residence times in storage tanks. These studies highlight the importance of tank design, location, and operation on water quality. Computer models, developed to explain some of the mixing and distribution issues associated with tank operation, were used to predict the effect of tank design and operation on various water quality parameters. Because of the diversity of the effects and the wide range of design and environmental conditions, the authors concluded that general design specifications for tanks are unlikely. They also concluded that models will most likely be refined and developed to facilitate site-specific analysis.

The impact of storage tanks on water quality was first noted by the authors during a study conducted by the US Environmental Protection Agency (USEPA) in conjunction with the South Central Connecticut Regional Water Authority (SCCRWA) (Clark et al. 1991). These studies will be used as the basis for additional development in this chapter.

Case Study Area

The town of Cheshire, Conn., served by SCCRWA, was the service area studied. Primarily a residential area, Cheshire is fed by two separate well fields, the North Cheshire well field, composed of four wells with a combined capacity of 3.6 mgd (13.6 ML/d), and the South well field, with two wells and a capacity of 2.5 mgd (9.5 ML/d). Storage is provided by two adjacent tanks, Prospect tanks 1 and 2, which "float on the system" (i.e., tanks that have little water movement and are designed to maintain system pressure), with 2.5 mil gal (9.5 ML) of storage each. The height of these tanks are 25 ft (7.5 m) and typically the tanks operate in a range of 14 to 19 ft (4.2 to 5.7 m). The operation of the wells is manually controlled by an operator who responds to the tank water levels. Average daily sendout for Cheshire is approximately 2.2 mgd (8.3 ML/d) during the winter and slightly more in the summer. Water use does not vary significantly by day of the week. Water use is highest during the morning and evening, and lowest during the night.

FIELD SAMPLING PROGRAM

As part of the overall USEPA-sponsored project, an extensive field sampling program was performed in the Cheshire system during November and December 1989. During this period, the fluoride feed into the water drawn from the two well fields was stopped for a seven-day period during which extensive sampling occurred throughout the system, including the water entering and leaving the Prospect tanks. The fluoride feed was later restarted and sampling was performed for the next seven-day period. Complete details on the sampling program, described in Chapter 8, are provided by Clark et al. (1991) and Skov, Hess, and Smith (1991). The results and analysis presented in this section relate only to the impacts and operation of the Prospect tanks.

A diagram showing the Cheshire distribution system, wells, the Prospect tanks, and the sampling sites is presented in Figure 9-1. Initially during the sampling study, the wells were operated to minimize the variation of water level within the Prospect tanks. After two days, this policy was changed to allow a much greater variation of water level in the tanks. This policy was followed for the remainder of the study. The well pumpage record and tank water level elevations during the study are graphed in Figure 9-2.

During the study, water entering and leaving the Prospect tanks was monitored on a frequent basis. An automatic sampler designed to sample at a set

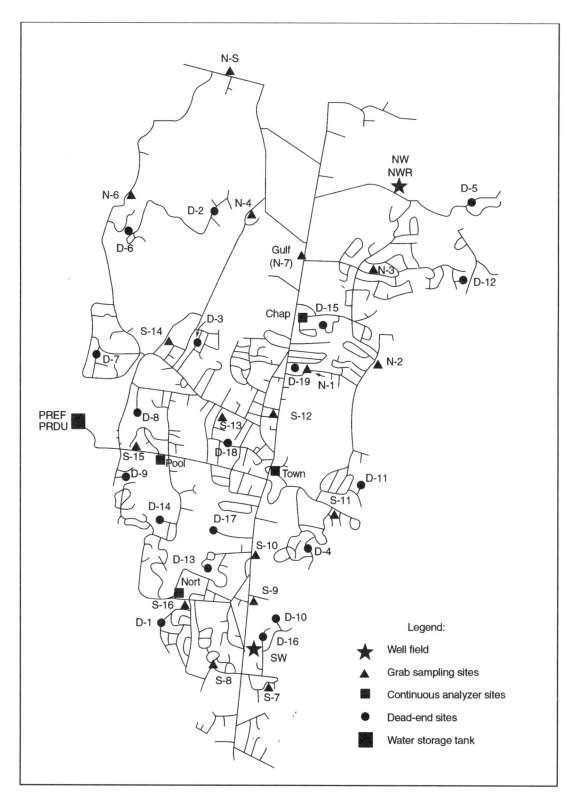

Figure 9-1 Cheshire service area map

interval (generally every one or two hours) was attached to the outlet line from the tanks. A continuous chart recorder monitored the flow in the combined inlet–outlet line. The resulting water quality profiles are shown in Figure 9-3. As illustrated, the variation between fluoride concentration levels entering the tank and leaving the tank were very significant. Only after about seven days did the levels of fluoride leaving the tank approach the levels entering the tank.

From this information, one could conclude that it would require about 10 days for the tank to reach an equilibrium. A second significant conclusion can be reached by examining the fluoride traces in the tank effluent. If the tank was a completely mixed reactor, the effluent-traces would be horizontal lines. As illustrated in Figure 9-3, during the period when the fluoride concentration levels in the tank exceeded the concentration levels in the system (i.e., during the first seven days of the sampling period when the fluoride was shut off), for each period of discharge, fluoride concentrations increased from a value representative of the inflow concentration to a value representative of the average concentration in the tank. This showed that the tank is not a completely mixed reactor and there was considerable short circuiting.

A similar inverse relation was observed during the second seven-day period after the fluoride had been turned back on and the level of fluoride concentration in the system exceeded the concentration level in the tank. Dur-

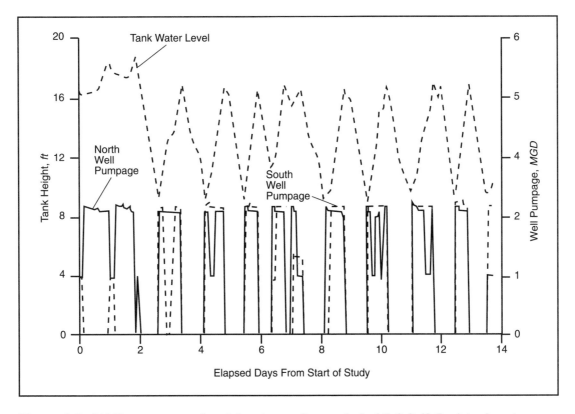

Figure 9-2   Well pumpage and tank level sampling period with 9-ft (2.7-m) tank water level variation

ing each period of tank discharge, it required approximately 8 hours for the effluent fluoride concentration to reach a constant concentration level, indicating a significant amount of short circuiting.

EFFECTS OF TANK CHARACTERISTICS ON WATER QUALITY

The field study showed that storage facilities could have a significant impact on water quality in a distribution system. In order to further investigate the effects of tank location and operation on the water quality in the system, a series of simulations were performed using a hydraulic and water quality model of the Cheshire system.

In the simulations, the effects of the location and operation of the tank were studied (Grayman, Clark, and Goodrich 1991). Location was studied by simulating the system with the tank located at its actual site (Prospect) and alternatively a tank of similar characteristics located in the vicinity of the North well field and near the center of town. These alternate locations are also shown in Figure 9-4. Various operations were simulated by varying the water level elevations in the tank that would activate the wells. A 9-ft (2.7-m) variation (well pumps on when the tank level drops to 9 ft [2.7 m] and off when it reaches

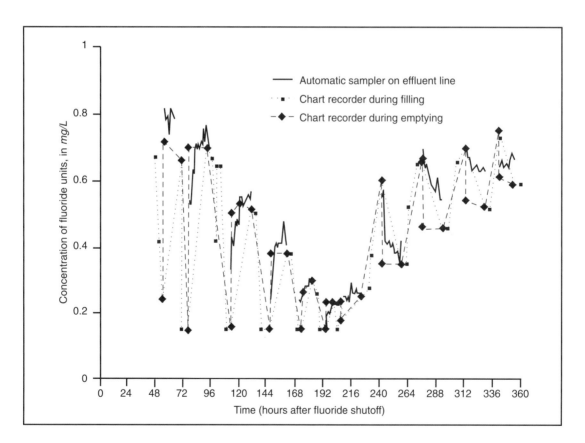

Figure 9-3 Fluoride concentrations at Prospect tank

18 ft [5.4 m]), a 6-ft (1.8-m) variation (on at 12 ft [3.6 m] and off at 18 ft [5.4 m]), a 3-ft (1.8-m) variation (on at 15 ft [4.5 m] and off at 18 ft [5.4 m]), and on a limited basis, a 1-ft (0.3-m) variation (on at 17 ft [5.1 m] and off at 18 ft [5.4 m]) were simulated.

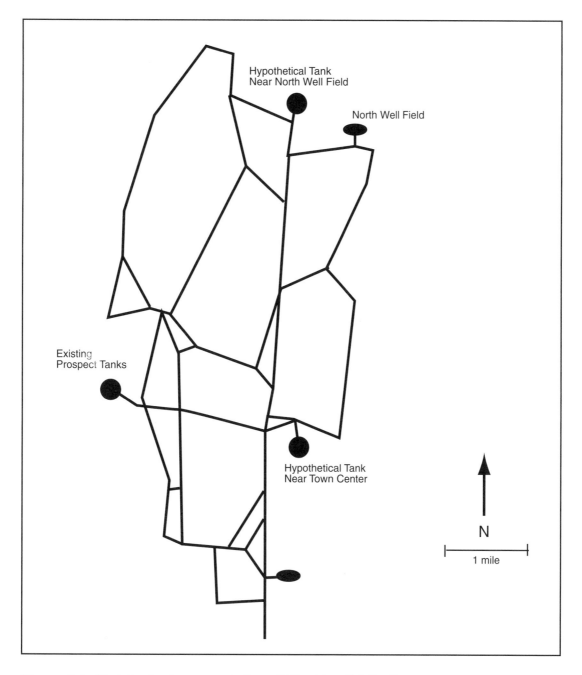

Figure 9-4 Skeletonized representation of Cheshire distribution system

SIMULATING WATER AGE

Because chlorine reacts with the natural organic matter (NOM) in water, water age is a key factor in the deterioration of water quality. With age, chlorine residuals will decrease and disinfection by-product (DBP) concentrations will increase. Therefore, age serves as a good indicator of water quality.

When simulating water age, an initial age of zero was assumed at all nodes and at the tanks. Water entering the system from wells was considered to be new water (age = 0). For the Prospect location, water level variations of 1, 3, 6, and 9 ft (0.3, 0.9, 1.8, and 2.7 m) were simulated. For the other locations, the minimum water level variation that was examined was 3 ft (1.8 m).

This analysis indicates a significant variation in age based on the different water level variations. For example, the long-term average water age in the Prospect tank is approximately 7 days when the water level in the tank is allowed to vary by 9 ft (2.7 m). However, when the variation in water level is reduced to 1 ft (0.3 m), the long-term average age increases almost threefold to approximately 20 days. A similar pattern is found for the alternate tank location near the town center. For the tank location near the North well, the difference in age is less for the different operating procedures.

This analysis shows the relatively close agreement in results for the Prospect and town center sites, with a significantly lower age when the tank is located in the vicinity of the North well. This phenomena is easily explained by the relatively short travel time from the well to the North well tank and, as a result, the infusion of primarily new water directly from the well during the time when the tank is filling.

As the water level variation decreases, the spread of older water through the system decreases and the age of the water in the tank and in the vicinity of the tank increases. For example, with a 1-ft (0.3-m) variation in the tank, the age of the water in the vicinity of the tank exceeds 18 days, while with a 9-ft (2.7-m) variation, the age of the water in this area is approximately 7 days. However, with the 1-ft (0.3-m) variation, for 32 of the 48 nodes, the maximum age of the water is less than 2 days, while for the 9-ft (2.7-m) variation, only 13 of the 48 nodes receive water with a maximum age less than 2 days. Thus, when the tank is located at Prospect, one must select between the two objectives of minimizing the age of the water in the vicinity of the tank or minimizing the spread of "old" water through the system.

When the tank is moved to the vicinity of the North well, the results are essentially the opposite. In this configuration, the spread of "old" water is greatest with the 3-ft (0.9-m) variation and the general age of the water throughout the system is largest (with the exception of the area in the vicinity of the South well, which receives almost all of its flow in the form of "new" water directly from the South well). When the tank is moved to the town center area, a different pattern results with the 6-ft (1.8-m) variation scenario resulting in the generally youngest water in the vicinity of the tank and throughout the system.

The large variation in results for the various tank locations and operating policies suggests that the age of water is very sensitive to both variables and that modeling is required to assess the likely response of the system to changes in tank location and operation. When water age is coupled with a time varying substance, such as chlorine residual, the effect of storage on water quality is very apparent.

CHLORINE RESIDUAL

A time varying parameter investigated in this study was chlorine residual. Chlorine residual was represented in the model by a first-order decay function. In this study, a value of 1 day^{-1} was used for the first-order decay coefficient. This value is somewhat larger than the observed decay in this distribution system but was useful for illustrative purposes (Characklis 1988). With a decay value of 1, chlorine residual has a half-life (will decay to half of its original value) of approximately 0.7 days. Chlorine residual is an excellent parameter for studying the effects of tanks because it combines the characteristics of a conservative constituent with the effects of age, and because of its real-life effects on the potential quality of water as it is delivered to the customer.

In the simulation, a chlorine residual value of 100 (corresponding to 100 percent of the chlorine residual in sources) was assumed at each source. Initially, all nodes (including the tank) were assumed to start with a chlorine residual of zero. The system was then simulated for a period until it reached a dynamic equilibrium (generally less than 10 days) and then the variation in chlorine residual throughout the system over the last 48-hour period was noted.

The minimum chlorine residual at each node for the different tank locations and tank water level variations were analyzed. The effects of water level variation at the Prospect tank are shown in Figure 9-5 for variations of 1, 3, 6, and 9 ft (0.3, 0.9, 1.8, and 2.7 m), respectively. As with the water age simulations, these effects are very apparent and very significant. As the water level variation increases, the spread of older water through the system increases and results in water with lower chlorine residual throughout the system. For example, with a 1-ft (0.3-m) variation in the tank, the chlorine residual is less than 20 percent (of the chlorine residual in the sources) for only 16 of the 48 nodes in the system. However, when the water level variation is increased to 9 ft (2.7 m), 38 of the 48 nodes have less than 20 percent of the source water chlorine residual. When the tank is moved to the vicinity of the North well, the results are much less dramatic. In this configuration, most of the system receives water that has been stored in the North well tank and consequently has a low chlorine residual for all operating procedures, with the exception of the area in the vicinity of the South well, which receives most of its flow in the form of "new" water directly from the South well.

Developing a Storage Tank Model

As indicated earlier, most water quality simulations assume complete mixing of the storage tanks examined. However, sampling data from the Prospect tank and other studies (Kennedy et al. 1991; Kennedy, Moegling, and Suravallop 1993) indicated that is not necessarily the case.

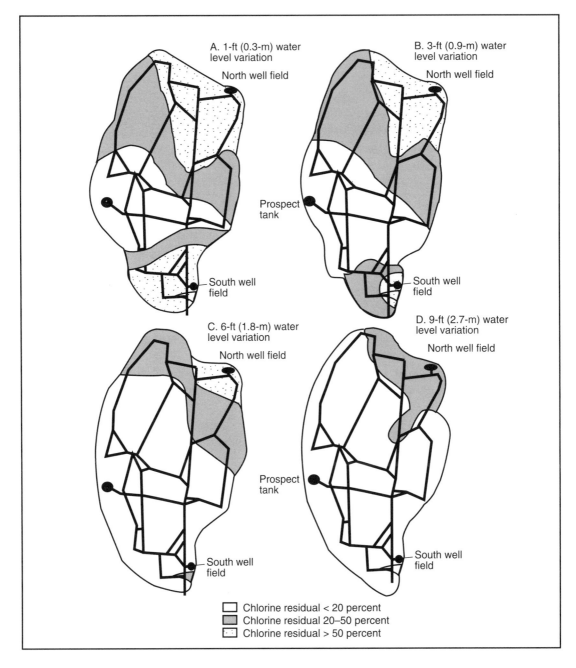

Figure 9-5   Minimum chlorine residual at Prospect tank

Grayman and Clark (1993) explored the use of compartment modeling to describe tank mixing and to deal with the nonuniform mixing in tanks, as described in the following section.

THE THREE-COMPARTMENT MODEL

Water quality models generally represent tanks as continuously stirred tank reactors (CSTR) (Grayman and Clark 1991; Grayman and Clark 1993). However, analysis of field sampling studies and data from laboratory studies of tanks indicates that there is considerable variation in the behavior of tanks and, that in many cases, the assumption that tanks are completely mixed is not a good representation.

Two related phenomena that lead to the breakdown of the CSTR assumption are short-circuiting between the inlet and outlet of a tank, and the presence of "dead areas" within the tank where mixing with the remainder of the tank is limited. In short-circuiting, a direct flow path between the influent and effluent line results in a last in–first out situation rather than full mixing. These phenomena may be due to design features or may result from environmental factors, such as temperature stratification.

A three-compartment model is suggested as a means of representing a wide range of tank configurations. As illustrated in Figure 9-6, the three-compartment model includes (1) compartment A, which represents the volume of the tank near the inlet/outlet where short-circuiting would most likely occur; (2) compartment C, which represents the tank volume where exchange

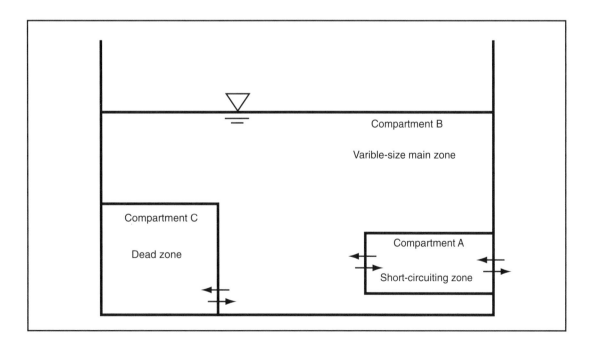

Figure 9-6   Schematic representation of three-compartment tank model

with the main body of the tank is limited (dead zone); and (3) compartment B, the main body of the tank.

These separate compartments are purely conceptual in nature. For example, compartment C may represent several different parts of the tank (e.g., corners in a rectangular basin or the upper portions of tank) where the interchange is limited.

In the model, compartments A and C are assumed to be of fixed volume. Compartment B varies in size as the water level in the tank changes. Each of the compartments is assumed to be represented as continuously stirred tank reactors, though at any time the concentration of the constituent of interest may vary between the compartments. A time lag is introduced through a numerical-difference solution technique that operates on an hourly time step.

In this model, all flow is assumed to enter and exit through compartment A. In order to maintain the fixed volume of this compartment, the flow rate between compartments A and B must be the same as flow rate between compartment A and the distribution system. The flow rate between compartments B and C is assumed to be adjustable. The lower the exchange rate, the "deader" the zone.

The control variables are the fixed volumes of compartment A and C, and the exchange rate between compartments B and C. If the volumes of compartment A and C are set close to zero, then the resulting model acts like a single-compartment CSTR. The values for the three input parameters and the relationship of the compartment volumes to the total volume of the tank results in highly variable behavior. It is important that the sum of the fixed volumes of compartments A and C is always less than the volume of the water in the tank or else the volume of compartment B would become negative and continuity would not be maintained. Similarly, the exchange rate between B and C should be such that the total volume exchanged during a time step is less than the volume of compartment C, otherwise questionable concentrations may result.

In the model, a constituent may be characterized as *conservative* or *nonconservative*. For nonconservative constituents such as chlorine, a first-order exponential decay is assumed.

The model may also be used to model the age of water. In this case, the age of water entering the tank must be estimated and the model then calculates the age in each of the three compartments.

This model has been implemented on a personal computer (PC) using the C language. Input to the model includes the three model parameters, the decay rate, hourly information on inflow and outflow rates for the tank, inflow concentrations, and initial concentrations by compartment. Historical data on the inflow and outflow concentrations for the period being simulated may also be provided by the user so that the model can display the comparison of simulated and historical concentrations. Output from the model includes both graphical and tabular displays. Graphical output includes a plot of predicted concentration over time for the three compartments and a plot of tank flow rates over the same time period. Figure 9-7 shows an example of such a plot. A second set of plots for each model run display the predicted concentrations for compartment A along with the historical data and a similar plot for compartment B, as shown in Figure 9-7.

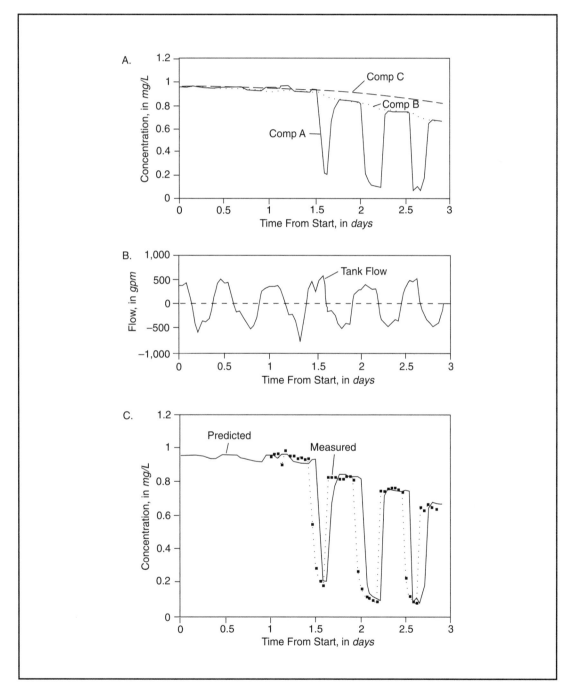

Figure 9-7   Sample output from three-compartment tank model

MODEL RESULTS

The three-compartment model was applied to both the Prospect tank and the Brushy Plains tank. Fluoride was modeled at the Prospect tanks, and fluoride, chlorine, and water age were studied at the Brushy Plains tank. For the Prospect tanks, there was no indication of a dead zone, so only compartments A and B were represented in the model.

As a basis of comparison, a single-compartment model was assumed and the results compared to the concentrations in the effluent line as determined from the field study (Figure 9-8). As illustrated, though the predicted concentration curve generally followed the trend of observed concentrations, it did not represent the variations that occurred for an 8- to 12-hour period as the tank switched from filling to emptying. As illustrated in Figure 9-8, the two-compartment model with compartment A equal to 100,000 ft^3 (2,800 m^3) resulted in a significantly improved representation of outflow concentrations.

The dominant phenomenon for the Brushy Plains tank was the apparent dead zone. To model this, compartment C volume was set at 100,000 gal (378,500 L) and an exchange rate between compartments B and C was set at 50 gpm (189 L/min). These parameter values resulted in a good fit when modeling fluoride concentrations in the tank discharge.

For chlorine, the decay coefficient also significantly affected model results, as shown in Figure 9-9, where a coefficient of K = 0 and K = 1 days^{-1}, respectively, were used.

Water age was also simulated for the Brushy Plains tank. Age of water was estimated for the tank inflow and for the initial tank water age by using historical chlorine data. Using these values, the water age was manually calculated and entered into the model. The resulting predicted time patterns of water age within each compartment is shown in Figure 9-10. The figure indicates a general increasing trend for water age in the tank, with fluctuation around this trend being most pronounced for compartment A as the tank changes from filling to emptying and vice versa. Because of the reduced exchange between compartments B and C, the age in compartment C is always greater than in compartment B. The relationship between water age in the three compartments is a good indicator of the dynamics occurring within the three-compartment tank model.

Parameterization of the model (i.e., selection of the three parameters) as described to this point, is a trial-and-error process. In addition to tank design, environmental parameters such as the relationship between ambient and water temperature, and conditions within the tank, may affect the model parameters. For the application to the two tanks, several different combinations of parameters were explored and even though an exhaustive search was not performed, an excellent relationship between simulated and actual values was achieved. It is hoped that greater experience with the model will lead to some general guidelines for selection of parameters to represent tanks of differing design or environmental conditions.

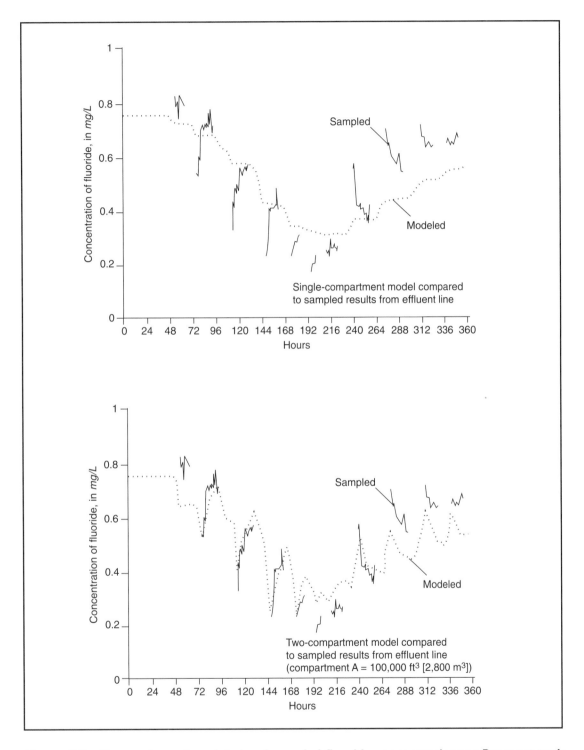

Figure 9-8 Comparison of modeled and sampled fluoride concentrations at Prospect tank

A consequence of long residence times is shown in Figure 9-11. As can be seen, the discharged water has very low residuals due to the long residence times in the tank (as implied from the previous study).

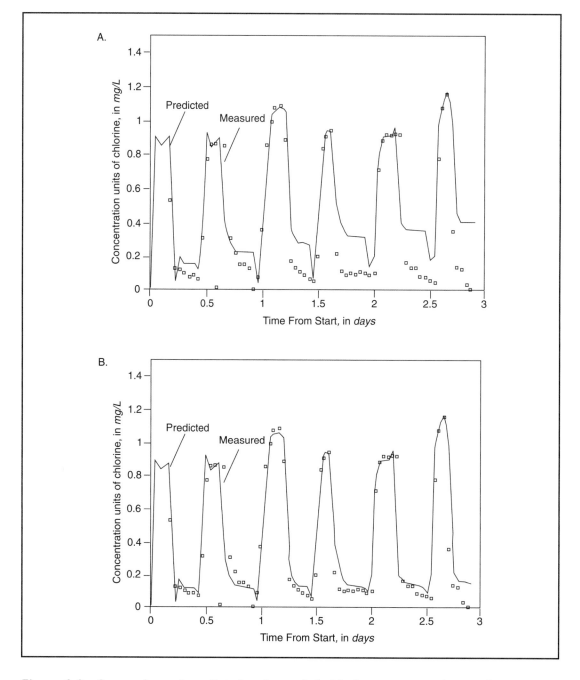

Figure 9-9 Comparison of predicted and sampled chlorine concentrations at Brushy Plains tank with A. no decay and B. decay

In the following sections, a more formal approach to compartmentalization is assumed. In addition, a technique that deals with time varying influent and effluent concentrations of a tracer is presented.

Mathematical Model Development

In this section, compartment models will be explored to characterize mixing in tanks. For purposes of model development, two types of tanks are considered. One is the traditional inflow and outflow storage tank, in which the water volume in the tank expands or contracts, depending on service demands. The Prospect and Brushy Plains tanks mentioned previously are examples of this type of tank. The other type of tank is illustrated by a tank located at the water treatment plant in Azusa, Calif. It functions as a flow-balancing mechanism and is used to impart contact times (CT) to the water. Water flows in and out of the tank simultaneously.

Mass balance equations will be used to describe the mixing conditions in each of the tank models. Compartments will be assumed to represent different

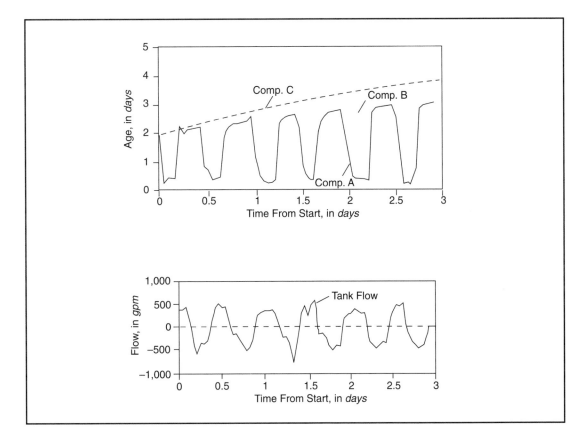

Figure 9-10   Simulated water age by compartment at Brushy Plains tank

mixing conditions in the various segments that make up the water volume in each tank (Mau et al. 1995; Grayman and Clark 1993).

For the traditional inflow and outflow tanks, the inlet and outlet pipes are near the bottom of the tank and, therefore, water entering the bottom of the tank will remain at the bottom displacing older water present at the beginning of the filling period. This general flow regime might be characterized as "first in, first out."

The approach to compartmentalization discussed in this paper is very similar to the approach suggested by Mau et al. (1995). However, Mau assumes steady state conditions for each inflow period (fluoride and inflow rate) and for each outflow period (outflow rate). In this work, the steady state assumptions have been relaxed and the actual time varying values have been approximated by polynomials.

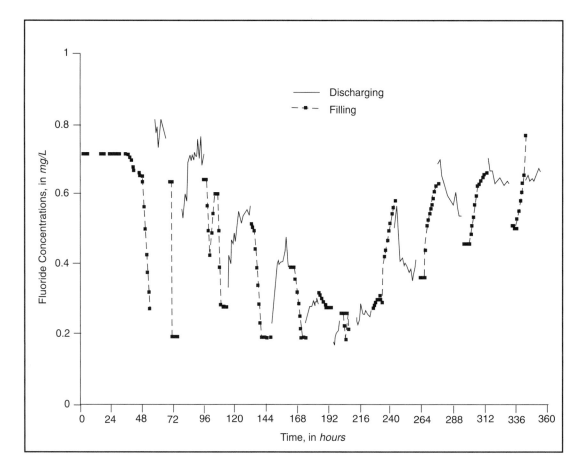

Figure 9-11 Fluoride concentration at Prospect tank

Tank Mixing Models

A compartmentalization approach will be used to represent different mixing zones in the water column within a tank. For purposes of this analysis, a maximum of three compartments is assumed. Figure 9-12 illustrates the inflow and outflow compartmental configurations associated with classical inflow–outflow tank systems. One- (completely mixed), two-, and three-compartment models are developed for the inflow–outflow and continuously flowing tank in the following sections (Clark et al. 1996).

In all the different configurations in this study, there are some basic formulas common to all of them. The general formula for the change in concentration of fluoride in the tank is

$$\frac{d(C_T V_T)}{dt} = F \cdot C \qquad \text{(Eq 9-1)}$$

Where:

F = flow, in ft/sec (m/sec)

C = concentration, in mg/L

d = differential

C_T = concentration of fluoride in tank, in mg/L

V_T = volume of tank, in ft^3 (m^3)

dt = rate of change in time

or equivalently,

$$V_T \frac{dC_T}{dt} = C_T \frac{dV_T}{dt} = F \cdot C \qquad \text{(Eq 9-2)}$$

Change in the water volume in the tank is described by

$$\frac{dV_T}{dt} = F \qquad \text{(Eq 9-3)}$$

Solving Eq 9-3 numerically, using Euler's method, we obtain:

$$V_T^{(n+1)} = V_T^{(n)} + h \cdot F^{(n)} \qquad \text{(Eq 9-4)}$$

Where:

h = integration step size

$n+1$ = the state of the "next" iteration

and other variables are as defined previously.

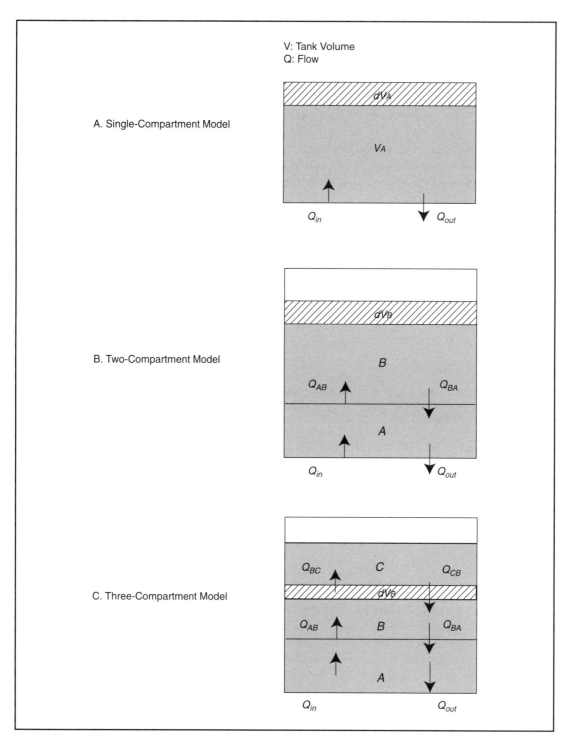

Figure 9-12   Compartment model configurations

INFLOW–OUTFLOW TANK MODEL

The inflow–outflow model is somewhat more complex so the equations describing one-, two-, and three-compartment models are contained in appendix A (Clark et al. 1996).

MODEL VALIDATION

Data from field studies conducted at the SCCRWA and the city of Azusa were used to validate the models discussed previously. Each of the systems and related field studies are described below.

SOUTH CENTRAL CONNECTICUT REGIONAL WATER AUTHORITY

The SCCRWA is a large regional water system serving more than 100,000 customers in the New Haven, Conn., region. The 16 service areas in eight pressure zones are fed by both surface water and groundwater sources (Clark et al. 1991). Though most of the system is tied together into two interconnected systems, the Cheshire area in the northern portion of the system is largely independent and was used as the basis for the study described.

The town of Cheshire is primarily a residential area. This service area is fed by two separate well fields. The North Cheshire well field is composed of four wells with a combined capacity of 3.6 mgd (13.6 ML/d), and the South well field has two wells and a capacity of 2.5 mgd (9.5 ML/d). Storage is provided by two adjacent tanks, Prospect tanks 1 and 2, which float on the system. Each has a capacity of 2.5 mil gal (9.5 ML). The height of these tanks is 25 ft (7.5 m), and typically the tanks operate in a range of 14 to 19 ft (4.2 to 5.7 m). The operation of the wells is manually controlled by an operator who responds to the water levels in the tanks. Average daily demand is approximately 2.2 mgd (8.3 ML/d) during the winter and slightly more in the summer. Water use does not vary significantly by day of the week. The highest water use occurs during the morning and evening, and the lowest water use occurs during the night.

Water entering and leaving the Prospect tanks was monitored frequently. An automatic sampler designed to operate at a set interval (generally every one or two hours) was attached to the outlet line from the tanks. A continuous chart recorder monitored the flow in the combined inlet–outlet line. The variation between fluoride concentration levels entering and leaving the tanks was very significant. Only after about seven days did the levels of fluoride leaving the tanks approach the concentration levels entering the tanks. From this information, it was concluded that it would require about 10 days for the tanks to reach an equilibrium.

The Cherry Hill/Brushy Plains service area covers approximately 2 mi^2 (5 km^2) in the town of Branford in the eastern portion of the SCCRWA area. This service area is almost entirely residential, containing single-family homes, apartments, and condominium units. Average water use during the sampling period was 0.461 mgd (1.7 ML/d). The distribution system is composed of 8- and 12-in. (203- and 304-mm) mains. The terrain in the service area is generally moderately sloping, with elevations varying from approximately 50 ft (15 m) mean sea level (MSL) to 230 ft (72 m).

Water is pumped from the Saltonstall system into the service area by the Cherry Hill pump station. Within the Cherry Hill/Brushy Plains service area, storage is provided by the Brushy Plains tank. The pump station contains two 4-in. (101-mm) centrifugal pumps with a total capacity of 1.4 mgd (5.3 ML/d). The operation of the pumps is controlled by water elevation in the Brushy Plains tank. Built in 1957, the tank has a capacity of 1 mil gal (3.7 ML). It has a diameter of 50 ft (15 m), a bottom elevation of 193 ft (58 m) (MSL), and a height (to the overflow) of 70 ft (21 m). During normal conditions, the pumps are set to operate when the water elevation in the tank drops to 56 ft (16.8 m) and to switch off when the elevation reaches 65 ft (19.5 m).

CITY OF AZUSA

The Azusa Light and Water Department (ALWD) sells water to portions of the San Gabriel Valley in Southern California. The department serves water to approximately 25,000 connections comprising an estimated population of 80,000. In 1993, the ALWD supplied 6.4 bil gal (24 GL) of water to its customers from a combination of treated surface water and groundwater from wells (Boulos et al. 1996).

The Ed Heck Reservoir was recently constructed in the vicinity of the Canyon Filtration Plant in Azusa. The aboveground, circular reservoir is constructed of prestressed concrete. The interior diameter of the reservoir is 154 ft (46.2 m) and the height is 30 ft (9 m). The reservoir holds roughly 4 mil gal (15 ML) when full. It is anticipated that the water level will normally be held between 15 and 27 ft (4.5 to 8.1 m) above the reservoir bottom.

A 24-in. (610-mm) diameter inlet pipeline from the Canyon Filtration Plant pump house feeds the reservoir. This pipeline enters near the base and has a right-angle elbow at its end. This elbow directs water counterclockwise along the reservoir perimeter. The energy of the inflowing water is sufficient to cause a visible rotation of the reservoir water in the same, counterclockwise direction. The filtration plant is capable of producing 10 mgd (37.8 ML/d).

A 30-in. (762-mm) diameter outlet pipeline carries water to the distribution system. This pipeline also lies at the reservoir base, 5.25 ft (1.6 m) clockwise of the inlet pipe. Distribution system operational practice makes it highly unlikely that water would enter the reservoir from the outlet pipeline.

FLUORIDE INFLOW–OUTFLOW MODEL

A key feature in this analysis is to model for the changes in fluoride concentration in the influent and to model the inflow and outflow rates for the tanks. Best-fit polynomials (shown in Tables 9-1, 9-2, and 9-3) were used to describe these variations.

Brushy Plains Tank

The Brushy Plains tank operates on alternating filling and emptying periods. The initial volume in the tank was 800,000 gal (3 ML). Initial fluoride concentration was 0.95 mg/L. In the three-compartment configuration, the exchange rate between compartments B and C was assumed to be 10 percent of the inflow or outflow, depending on the filling or emptying period. Column 1 of Table 9-1 shows the inflow and outflow periods, column 2 contains the polynomial for the fluoride entering the tank, and columns 3 and 4 contain the models for inflow and outflow in gallons per minute. In the three-compartment model, the exchange rates between compartments B and C were assumed to be 10 percent of the inflow and outflow.

Prospect Tank

The Prospect tank is either filling, emptying, or sitting idle. The initial volume of the tank was 2,961,788 gal (11,210,367 L). The initial fluoride concentration was 0.75 mg/L. In the three-compartment model, the exchange rate between compartments B and C was assumed to be 10 percent of the inflow when filling, and 10 percent of outflow when emptying.

Table 9-1   Influent and effluent models for Brushy Plains tank

Time, in $hours$	Fluoride In, in mg/L	Inflow, in gpm	Outflow, in gpm
0–3	$0.06x - 0.00003x^3$	$0.07x^3 - 0.00008x^5$	—
3–9	—	—	$0.01x^3$
9–14	$7.3 - 0.2x$	$0.0003x^4$	—
14–21	—	—	$0.18x^2$
21–28	$82 - 3x + 0.03x^2$	$0.4x - 0.000002x^5$	—
28–34	—	—	$1597x - 42x^2 + 0.004x^4$
34–39	$194.5 - 6x + 0.05x^2$	$0.00003x^4$	—
39–46	—	—	$8(e - 13)x^8$

NOTES:

x = time in hours

e = base for natural (Napierian) logarithm

Table 9-2 contains the equation describing inflow and outflow for the Prospect tank. Figure 9-13 illustrates the use of influence models for the Prospect tank. The figure shows the influent flow data and model for hours 129–140, the actual and modeled fluoride concentration for hours 130–140, and the actual and modeled effluent flows for hours 140–150.

Table 9-2 Influent and effluent models for Prospect tank

Time, in *hours*	Fluoride In, in *mg/L*	Inflow, in *gpm*	Outflow, in *gpm*
0–21	$11 - 0.7x + 0.02x^2 - 0.0002x^3$	$789x - 49x^2 + 0.8x^3 - 7(e-5)x^5$	—
21–35	—	—	$8,385 - 3x^2 + 5(e-6)x^5$
35–43	$133 - 4x + 0.02x^2$	$3,106,962 - 81,385x + 11x^3 - 0.001x^5$	—
43–61	—	—	$-561,047 + 12,262x - x^3 + 4(e-5)x^5$
61–80	$2(e-6)x^3 - 2(e-10)x^5$	$0.14x^3 - 0.003x^4 + 1.2(e-5)x^5$	—
80–98	—	—	$2,732x - 44x^2 + 0.2x^3$
98–114	$16 - 0.002x^2 + (e-5)x^3$	$10,131,904 - 194,505x + 1,049x^2 - 0.01x^4$	—
114–129	—	—	$-653,258 + 8,373x - 27x^2$
129–140	$3 - 9(e-5)x^2$	$-779,239 + 9,128x - 27x^2$	—
140–150	—	—	$-1,158,074 + 12,994x - 36x^2$
150–162	$22 - 0.23x + 0.0006x^2$	$-91,658 + 6,790x^{1/2}$	—
162–167	—	—	$-1,851,281 + 14,050x - 0.1x^3$
167–174	$2(e-8)x^3$	$-0.0003x^3 + 2(e-6)x^4$	—
174–192	—	—	$-15,256,431 + 132,728x - 2x^3 + 1.2(e-5)x^5$
192–210	$-4.3 + 0.02x$	$-285,072 + 10x^2 - 0.00009x^4$	—
210–227	—	—	$-38,452,981 + 379,978x - 1,001x^2 + 6(e-6)x^5$
227–242	$-90 + 0.7x - 0.001x^2$	$-40,357,401 + 34,910x - 780x^2 + 3(e-11)x^7$	—
242–261	—	—	$-460,040 + 2,459x - 0.01x^3$
261–279	$-4.8 + 0.02x$	$-23,608,794 + 135,897x - 0.7x^3 + 1.2(e-11)x^7$	—
279–297	—	—	$-2,790,770 + 17,335x - 27x^2$
297–308	$242 - 1.5x + 0.002x^2$	$-4,051,835 + 24,001x - 36x^2$	—
308–321	—	—	$2(e-6)x^4 - 6(e-9)x^6$

Table 9-3 Influent and effluent models for Azusa tank

Time, in *hours*	Fluoride In, in *mg/L*	Outflow, in *gpm*	Inflow, in *gpm*
0–1.2		$10,298 - 15,818x + 11,628x^2$	
1.2–2.8		$47,259x - 63,913x^2 + 30,401x^3 - 4,788x^4$	
2.8–5	0.5	$13,421,959 - 9,094,333x + 2,302,515x^2 - 258,071x^3 + 10,805x^4$	
5–6.8		$13,421,959 - 9,094,333x + 2,302,515x^2 - 258,071x^3 + 10,805x^4$	
9–9.4	$-693 + 148x - 7.9x^2$		
9.4–10.2	$0.004x^4 - 0.0007x^5 + 3(e-5)x^6$	$89x^4 - 18x^5 + 0.92x^6$	
10.2–12.4	$0.05x^2 - 0.003x^3$	$37,298,901 - 8,032,286x + 107,453x^3 - 7,207x^4 + 7.7x^6$	
12.4–13		$1,472,133 - 225,809x + 8,689x^2$	
13–13.6			
13.6–14.8		$1,677,445 - 235,757x + 8,309x^2$	
14.8–16	$0.002x^3 - 0.00008x^4$	$1,757,327 - 228,032x + 7,419x^2$	
16–17		$2,990,031 - 362,294x + 10,993x^2$	
17–18.2		$2,197,368 - 248,943x + 7,069x^2$	
18.2–19		$8,308,758 - 891,234x + 23,908x^2$	
19–20.2		$2,985,626 - 306,539x + 7,880x^2$	$5,923 - 0.7x^3 + 0.09x^4 - 0.004x^5 + 9(e-5)x^6 - 1(e-6)x^7 + 4(e-9)x^8$
20.2–21		$7,260,595 - 704,612x + 17,108x^2$	
21–22	$-283 + 48x - 3x^2 + 0.09x^3 - 0.0009x^4$	$5,194,822 - 481,960x + 11,191x^2$	
22–24.6		$1(e-6)x^8 - 4(e-8)x^9$	
24.6–26		$0.0000008x^7$	
26–28		$50,158,565 - 4,949,812x + 137,361x^2 - 31x^4$	
28–30	$0.0002x^3 - 0.000005x^4$	$0.008x^4$	
30–31		$10,793,247 - 707,438x + 11,598x^2$	
31–32	$1(e-5)x^4 - 3(e-7)x^5$	$12,431,154 - 788,490x + 12,509x^2$	
32–34		$13,004,943 - 799,565x + 12,295x^2$	
34–35.8		$2(e-6)x^7 - 5(e-8)x^8$	
35.8–37.2		$-0.0013x^5 + 0.00004x^6$	
37.2–39.2		$353x^2 - 0.02x^5 + 6(e-6)x^7$	
39.2–40.2		$21,820,731 - 1,099,327x + 13,848x^2$	
40.2–41.2	0.5	$23,337,896 - 1,147,050x + 14,096x^2$	
41.2–42.2		$23,020,403 - 1,104,494x + 13,250x^2$	
42.2–43		$37,758,587 - 1,772,876x + 20,812x^2$	
43–44		$2,177,644 - 1,005,085x + 11,546x^2$	

table continues next page

Table 9-3   Influent and effluent models for the Azusa tank (continued)

Time, in hours	Fluoride In, in mg/L	Outflow, in gpm	Inflow, in gpm
44–44.8		$48,100,956 - 2,165,256x + 24,368x^2$	
44.8–45.8		$25,101,748 - 1,107,189x + 12,210x^2$	
45.8–46.8		$24,728,874 - 1,068,258x + 11,538x^2$	
46.8–47.8		$27,900,434 - 1,182,419x + 12,529x^2$	
47.8–48.8		$22,150,451 - 915,628x + 9,464x^2$	
48.8–50.4		$12,772,259 - 513,661x + 5,165x^2$	
50.4–52.6	0.5	$1(e - 10)x^8$	$5,923 - 0.7x^3 + 0.09x^4$ $- 0.004x^5 + 9(e - 5)x^6$ $- 1(e - 6)x^7 + 4(e - 9)x^8$
52.6–53.8		$5,923 - 0.7x^3 + 0.9x^4 - 0.004x^5$ $+ 9(e - 5)x^6 - 1(e - 6)x^7 + 4(e - 9)x^8$	
53.8–54.6		$4x^4 - 0.12x^5 + 1.4(e - 5)x^7$	
54.6–55.6		$22,223,081 - 604,964x + 66x^3$	
55.6–56.6		$9.3(e - 13)x^9$	
56.6–57.4		$2.6(e - 9)x^7$	
57.4–59.2		$0.0005x^4$	
59.2–60.8		$2(e - 9)$	

Azusa Tank

The Azusa tank operates on a simultaneous filling and emptying mode. The initial volume was 3 mil gal (11.3 ML) with an initial fluoride concentration of 0.5 mg/L. Table 9-3 contains the models for each time period for the Azusa Tank.

Model Application

An objective of this analysis was to determine which compartment assumption best described the mixing conditions for the three tanks. A variable was the percentage of initial volume for each compartment in a given compartment model. For example, for the two-compartment model in the Prospect tank, the assumptions ranged from 10 percent for the first compartment and 90 percent for the second compartment, to 70 percent for the first compartment and 30 percent for the second compartment. Calculated outcomes were regressed against the observed data from the effluent data and the respective goodness of fit measurements (R^2) were calculated. The differences between the means predicted by each model were compared to determine if they differed from one another. If they differed, then the model with the highest R^2 was assumed to be the best model. If they did not differ, then the simplest model was assumed to be the best model. Results from each tank are discussed below.

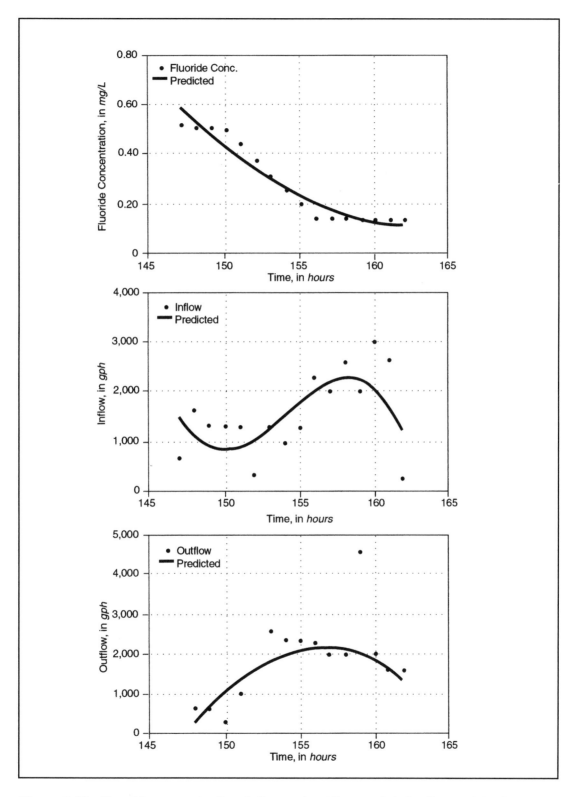

Figure 9-13 Fluoride concentration, inflow and outflow models for Prospect tank

PROSPECT TANK

Table 9-4 contains the results of the analyses for the Prospect tank. The one-compartment model's best-fit test resulted in a $R^2 = 70$ percent. For the two-compartment model, several assumptions for the variable and fixed compartments were considered. The solution started with 10 percent of the initial volume for the fixed part and 90 percent for the variable part. This amount was then increased by 10 percent for the fixed and decreased by 10 percent for the variable compartment assumption. The 10 percent and 20 percent partitioning also did not result in a better fit than the one-compartment model, but the fit for the 30 percent (30–70) partition improved and yielded an $R^2 = 82.6$ percent. The 40 percent (40–60) assumption had a very similar R^2 of 82.7 percent, but was not a major improvement over the 30 percent assumption.

Three-compartment models were run with 10 percent for each fixed and 80 percent for the variable compartment. This amount was changed to 20 percent for the fixed and 60 percent for the variable compartment. The R^2 for the first case was 72.9 percent. The best-fit test indicated that the three-compartment model for the 20–60–20 percent partitioning was the overall best fit with an R^2 of 83.9 percent.

For this case, the three-compartment model with a 20–60–20 portioning was assumed to be the best model. This same appraisal was used for each tank.

BRUSHY PLAINS

Table 9-5 contains results for the Brushy Plains tank. The one-compartment model yielded an R^2 of 91.4, which indicates that the one-compartment assumption was good. The two-compartment model was examined for various configurations, including 10–90 (i.e., 10 percent of the initial volume for the fixed part and 90 percent for the variable part), 20–80, 30–70, 40–60, 50–50, 60–40, and 70–30 partitioning. The 70–30 model resulted in the highest R^2 of 91.96 percent and was assumed to be the best model.

For the three-compartment models, partitioning ranged from 10–80–10 to 30–40–30 (i.e., 30 percent of the initial volume for the fixed compartments

Table 9-4   Best-fit results for Prospect tanks

Model		R^2
One-Compartment Model		70%
Two-Compartment Model	10–90%	68.7%
	20–80%	77.9%
	30–70%	82.6%
	40–60%	82.7%
Three-Compartment Model	10–80–10%	72.9%
	20–60–20%	83.9%

A and C and 40 percent for the variable compartment B). The highest R^2 was 89.6, which indicated poorer fit than the one- or two-compartment models.

AZUSA CITY TANK

Table 9-6 contains results from the Azusa City tank.

Although the one-compartment model resulted in a good fit, the overall best-fit model was the two-compartment model with 70–30 partitioning.

The best-fit test for the one-compartment model resulted in R^2 = 94 percent, which indicates that one-compartment model is a very suitable configuration for this tank.

Several different partitionings for the two-compartment model were examined. Partitions ranged from 10–90 to 60–40 for the fixed-variable compartments.

Table 9-5   Best-fit results for Brushy Plains tank

Model		R^2
One-Compartment Model		91.4%
Two-Compartment Model	10–90%	68.9%
	20–80%	84.4%
	30–70%	89.4%
	40–60%	90.9%
	50–50%	91.6%
	60–40%	91.9%
	70–30%	91.96%
Three-Compartment Model	10–80–10%	69.6%
	15–70–15%	79.8%
	20–60–20%	85.2%
	25–50–25%	88.2%
	30–40–30%	89.6%

Table 9-6   Best-fit results for Azusa City tank

Model		R^2
One-Compartment Model		94.3%
Two-Compartment Model	10–90%	90.0%
	20–80%	89.5%
	30–70%	88.7%
	40–60%	88.1%
Three-Compartment Model	10–80–10%	91.0%
	15–70–15%	90.0%

The 10–90 configuration that yielded an $R^2 = 90$ percent was the best among the two-compartment partitioning.

For the three-compartment model, only the following two cases were examined: (1) 10 percent for each fixed compartment and 80 percent for the variable compartment and (2) 15 percent for each fixed and 70 percent for the variable compartment. The first case resulted in the better $R^2 = 91$ percent. It was concluded that the one-compartment model was the best configuration for this tank. Refer to appendix A for equations for one-, two-, and three-compartment models.

References

AWWA. 1989. *AWWA Manual M1, Distribution System Requirements for Fire Protection.* Denver, Colo.: American Water Works Association.

Boulos, P.F., W.M. Grayman, R.W. Bowcock, J.W. Clapp, L.A. Rossman, R.M. Clark, R.A. Deininger, and A.K. Dhingra. 1996. Hydraulic Mixing and Free Chlorine Residual in Reservoirs. *Jour. AWWA*, 88(7):48–59.

Characklis, W.G. 1988. *Bacterial Regrowth in Distribution Systems.* Denver, Colo.: American Water Works Association Research Foundation and American Water Works Association.

Clark, R.M., W.M. Grayman, J.A. Goodrich, R.A. Deininger, and A.F. Hess. 1991. Field Testing Distribution Water Quality Models. *Jour. AWWA*, 83(7):67–75.

Clark, R.M., F. Abdesaken, P.F. Boulos, and R.E. Mau. 1996. Mixing in Distribution System Storage Tanks: Its Effects on Water Quality. *Jour. Environ. Eng.*, 122(9):814–821.

Danckwerts, P.V. 1958. Continuous Flow Systems. *Chemical Engineering Science*, 2(1):1–18.

Germeles, A.E. 1975. Forced Plumes and Mixing of Liquids in Tanks. *Jour. Fluid Mechanics*, Vol. 71, Part 3.

Grayman, W.M., R.M. Clark, and J.A. Goodrich. 1991. The Effects of Operation, Design and Location of Storage Tanks on the Water Quality in a Distribution System. In *Proc. of the Water Quality Modeling in Distribution Systems Conf.* Denver, Colo.: American Water Works Association Research Foundation and American Water Works Association.

Grayman, W.M., and R.M. Clark. 1991. Water Quality Modeling in a Distribution System. In *Proc. AWWA 1991 Annual Conference.* Denver, Colo.: American Water Works Association.

Grayman, W.M., and R.M. Clark. 1993. Using Computers to Determine the Effect of Storage on Water Quality. *Jour. AWWA*, 85(7):67–77.

Kennedy, M.S., S.S. Moegling, and K. Suravallop. 1993. Assessing the Effects of Storage Tank Design. *Jour. AWWA*, 85(7):78–88.

Kennedy, M.S., S. Sarikelle, S. Moegling, and K. Suravallop. 1991. Mixing Characteristics in Distribution System Storage Reservoirs. In. *Proc. AWWA 1991 Annual Conference.* Denver, Colo.: American Water Works Association.

Mau, R., P. Boulos, R. Clark, W. Grayman, R. Tekippe, and R. Trussell. 1995. Explicit Mathematical Models of Distribution System Storage Water Quality. *Jour. of Hyd. Eng.*, 121(10):699–709.

Skov, K.R., A.F. Hess, and D.B. Smith. 1991. Field Sampling Procedures for Calibration of a Water Distribution System Hydraulic Model. In *Water Quality Modeling in Distribution Systems.* Denver, Colo.: American Water Works Association Research Foundation/USEPA and American Water Works Association.

Getting Started in Modeling

Applying distribution system network modeling requires both an understanding of the underlying fluid mechanics associated with design and operation of the network, and engineering judgment derived from field experience. Although it is impossible to reduce the use of models to an all-encompassing set of rules and procedures, the following general steps can provide guidelines for those just beginning a modeling effort.

Steps in Modeling

1. *Definition of scope.* The first step in the modeling process is to define the general scope of the modeling effort. Will the model be used to investigate broad planning alternatives or will it be used to answer detailed design questions, such as those involved in determining pipe diameters to meet fire-fighting needs? Will the model be applied to a small town or to a major metropolitan area? Will water quality modeling be included as part of the effort? Does the model need to be integrated with an existing computer-aided drafting (CAD) or design package? These are only a few of the questions that should be posed in defining the overall scope of the modeling effort. Such questions should always be considered when the modeling process is initiated.

2. *Model selection.* The model is the tool that provides the computational capability to assess design policy alternatives associated with the network. There are many models available that may satisfy a user's needs and there are many factors that should be considered in the selection process. These factors are discussed later in this chapter.

3. *Model setup.* Once a model is chosen, the initial step is to install the model on the computer and to become familiar with its capabilities

and limitations. Actual installation is generally quite straightforward. This familiarization process may be accomplished through formal training courses or trial and error using example problems. However, before full-scale application of the model, the user should become confident in his or her ability to represent the network and to interpret results from the model.

4. *Network representation.* In all models, the distribution system network is represented by a series of links and nodes that correspond to pipes, junctions, pumps, valves, tanks, etc. Depending on the user's objectives, this may include a complete representation of the network or a selected number of primary components. Data may be entered manually, through various automated mechanisms involving digitization, or by scanning maps and integration with CAD packages or geographic information systems (GIS). A more complete discussion of the network representation process is provided later in this chapter.

5. *Calibration.* If there was a perfect representation of all the components that make up a system and a model that exactly represented the physical processes occurring in the system, then application of the model could commence after the network representation has been completed. However, there are always assumptions, uncertainties, and unknowns in representing the system, which require a calibration process. In the calibration step, various parameters in the network representation are adjusted so that the results of the model match observed field measurements. Various options for calibration are discussed later in this chapter.

6. *Validation.* In order to verify that the network representation and calibration process is adequate, the model is validated. In the validation process, model results are compared to a set of independent field observations. If significant anomalies or deviations between observed and modeled results occur, then refinements in the representation process or further calibration is warranted.

7. *Problem definition.* Once the validation step has been completed, the model can be used to achieve the objectives laid out in the initial problem definition (step 1). This may include such considerations as the input of a pattern of water usage corresponding to different seasons, investigation of an operating assumption for the distribution system, evaluation of source water quality effects, etc.

8. *Model application.* Full-scale application of the model may be performed. This may initially involve adjustment of some internal model parameters, such as time step assumptions, and then examination of multiple scenarios.

9. *Analysis and display.* Application of network models to various scenarios can generate large amounts of output. This output is used in

the calibration and validation process and to confirm that the model is performing in a reasonable manner. The output is then analyzed to determine how the system responds under the different scenarios to be considered. Many models contain graphic displays and statistical analysis tools or external software packages that may be used in this process.

Selecting a Model

Selecting a hydraulic–water quality model for a distribution system should be a process of matching the specific needs of the user to the available models.

SELECTION CRITERIA

Following is a list of criteria that should be considered when selecting a hydraulic–water quality model for a distribution system:

- Cost

- User interface

- Ease of use

- Components represented

- Water quality modeling characteristics

- Robustness

- Size constraints

- Support

- Documentation

- Speed

- Hardware requirements

- Database/GIS/CAD/SCADA[*] integration

- Flexibility

- Training

Each of these criteria are discussed in more detail in the following sections.

---

[*] Supervisory control and data acquisition.

Model Cost

Hydraulic–water quality models range in cost from free to about $100,000. In some cases, the cost of the model depends on the size of the distribution system that is to be modeled. Some models operate in conjunction with proprietary packages, such as CAD software, so these costs should also be considered when comparing software packages. For smaller water utilities or consultants, or when modeling is used infrequently, large expenditures for modeling software would generally not be considered prudent and therefore, the cost may be a predominant factor in model selection. However, for larger, more frequent applications, the capital cost associated with the model acquisition may represent only a small portion of the total budget allocated to distribution system studies. In this case, the initial cost of a model may only be a minor factor in model selection.

User Interface

The look and feel of a model is reflected in the user interface, or the mechanism by which the user interacts with the model. There have been tremendous advances in user interface development during recent years. Portions of a typical user interface (input and output) from a model in the 1980s are shown in Figure 10-1. By comparison, a graphical user interface (GUI) representative of the current state-of-the-art is illustrated in Figure 10-2. In a few short years, the GUI progressed from a modeling feature rarely found, to a feature that most users now expect. In water distribution system models, a GUI may include various graphical and tabular options for visualizing results, and graphical methods for entering or editing the characteristics of the distribution system network.

Currently there is a trend toward a Windows-based standard for GUIs. Selecting a user interface is a very personal and subjective process. Demo disks available for many systems provide one mechanism for examining the interface. A preferred method is to examine a full-featured trial package that allows the user to try the model out on small (e.g., 25-link) systems or for a limited time. Such trial packages are available for many models.

Ease of Use

Ease of use is another subjective factor that is important to many users. The trend is toward models that are relatively easy to use. On-line help, GUIs intuitive and standard designs all contribute to ease of use. The primary source of information on this topic is through other users. Since this topic is quite subjective, it is best to communicate with others whose general background and modeling level approximates yours.

A. Partial Input File

CHESHIRE SYSTEM—SCENARIO 5 BASED ON WMGSKL2A 6-18-89

1	1	2	4312	110
2	3	4	6209	110
3	5	6	6759	110

ELEVATION (node elevation from DEM)

1	115.7
2	210.5
3	185.1

OUTPUT

2	32.9775
5	40.08583
6	16.6325

TANK 49 22.

RATIO 1.496

ACCURACY .5 .5

END

OC

9

B. Partial Output File

WADISO—WATER DISTRIBUTION SYSTEM OPTIMIZATION
VERSION: MARCH 19, 1987
PUBLIC DOMAIN PROGRAM DEVELOPED BY GESSLER, SJOSTROM & WALSKI
US ARMY CORPS OF ENGINEERS—WATERWAYS EXPERIMENT STATION
MAX NODE NO = 900 NUMBER OF NODES = 700
MAX PIPE NO = 900 NUMBER OF PIPES = 700

PIPE NETWORK ANALYSIS AND OPTIMIZATION
JOB: CHESHIRE SYSTEM—SCENARIO 5 BASED ON WMGSKL2a 6-18-89

NODE DATA

NODE NO.	ELEV. FT.	OUTPUT GPM	E.G.L. FT.	PR. HEAD FT.	PRESSURE PSI	
1	115.7		430.0	314.3	136.2	
2	210.5	49.	423.0	212.5	92.1	
3	185.1		418.3	233.2	101.0	
4	136.4		419.3	282.9	122.6	
5	252.4	60.	414.9	162.5	70.4	
6	193.9	25.	415.0	221.1	95.8	
7	177.2	81.	415.2	238.0	103.1	
49	393.0	−6.	415.0	22.0	9.5	SUPPLY

PIPE NO.	NODES FROM TO		DIAM. IN.	LENGTH FT.	COEF	FLOW GPM	VEL. FT/SEC	HEAD LOSS
1	1	2	16.0	4312.0	110.	1404.0	2.2	6.9
2	4	3	16.0	6209.0	110.	408.	.7	1.0
3	6	5	12.0	6759.0	110.	44.	.1	.0
4	7	8	8.0	3684.0	80.	31.	.2	.3
5	9	10	8.0	5424.0	80.	156.	1.0	7.9
6	12	11	12.0	3262.0	110.	197.	.6	.6

Figure 10-1 **Sample file for typical model in the 1980s**

Components Represented

In general, a network model should include all primary components that make up the water distribution system. The primary differences among the various available models are in the representation of tanks, valves, pumps, and control statements. For example, some models allow the direct representation of tanks when the cross-sectional area varies with elevation, while others require a constant area. Some models do not explicitly represent the full variety of valves (pressure reducing, pressure sustaining, flow control, etc.) that are commonly used in distribution networks. Pump characteristics are generally represented by a pump curve (flow versus head), although there may be subtle differences in the exact method used by the model. *Control statements* refer to the way that the user can tell the model how components should operate under dynamic conditions. For example, all models should permit a pump to be turned on or off by a clock or the water level in a tank. Few models allow compound control statements in which a component is controlled by the state of multiple components (i.e., the use of "and/or" statements). This capability

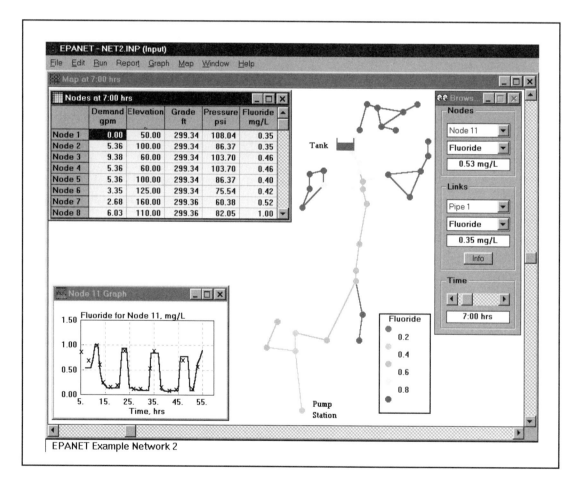

Figure 10-2 EPANET graphical user interface

might be critical in a specific allocation. When a system contains a variety of components or controls, the ability of the alternative models should be specifically checked before selecting a model for that particular use.

Water Quality Modeling Characteristics

Hydraulic–water quality models vary widely in water quality capabilities. Some models permit only steady state modeling (i.e., hydraulic and water quality flows and loads cannot vary with time), while others simulate the more widely applicable extended period simulation (EPS) modeling. Some variations in water quality modeling characteristics include the capability to model conservative substances, substances that decay (or grow) according to a first-order function, age of water, and coupled constituents (i.e., the rate of change in one constituent depends on the concentration of a second constituent). Chlorine residual kinetics is frequently of interest. Some models can represent only the chlorine bulk decay, while others can also represent chlorine consumption due to wall demand.

Robustness

Robustness is the ability of a model to find a correct solution under a wide range of situations. Since all dynamic water distribution system models use a numerical scheme, the convergence and stability of the model is important. Convergence problems occur most often in systems with multiple valves (i.e., pressure-reducing valves), in which each valve influences the operation of another valve. For such complex systems, a small-scale test to find which models handle these problems best is suggested prior to selection.

Size Constraints

Most models have some limitation on the number of links, nodes, and components that may be simulated. Such constraints may depend on the memory capacity of the computer being used. In some cases, size may not be an absolute constraint but large systems may result in unacceptably slow performance. The general trend in models is toward supporting simulation of larger and larger systems.

Support

Technical support may be a very important issue for the inexperienced modeler or modeler who is unfamiliar with a newly acquired modeling package. Support requirements may range from none to extensive telephone or on-site support. Continuing support may be part of the initial purchase price of the package or a separate annual support contract may be required. The type of support available may range from use of the model to problem definition and analysis. The quality of support may also vary among vendors and discussion of this issue with other model users is recommended.

Documentation

Virtually all models available for public use contain some form of documentation. The trend is toward emphasizing on-line help as a supplement or replacement for traditional printed material. A typical Windows help screen from the EPANET model is shown in Figure 10-3.

Speed

The speed of a model may be measured in terms of the computer time required to run an application of the model and/or the time to graphically display the results. Speed depends on the inherent speed of the computer, but with the trend toward faster computers, this has become less of an issue. If a relatively small system (i.e., less than 500 pipes) is being modeled in a steady state mode or for short periods (less than 24 hours), most models will run in a few minutes, thus speed may not be an overriding selection factor in any case. However, for larger systems and/or long simulations, run times may be measured in hours and the inherent speed of the model may then become a very important factor. Benchmark tests have been run independently and by model vendors and these results should be examined as part of the selection process.

Hardware Requirements

Most models operate on a standard personal computer (PC) with moderate random access memory (RAM) and speed. However, some models require more memory or greater speed than others for proper operation. Though the trend in the industry is toward more powerful machines, some models may require users to upgrade their present computers or to purchase new computers.

Database/GIS/CAD/SCADA Integration

Water distribution system models can be used to analyze design and operational scenarios. Frequently the information needed to apply the model or the results generated by the model come from or are used by external packages, such as database management systems, GIS, CAD, or SCADA. The model may be integrated with such packages through tight linkages or by sharing files. Some models operate from within a proprietary CAD package while others have a stand-alone CAD package provided with the modeling software. Such tight integration may provide for a very smooth interface but requires the user to understand the CAD package.

Input File Section: [TANKS]
Definition Sketch Examples

Purpose:
Describes each storage tank or reservoir in the network

Formats:
node elev (initlevel minlevel maxlevel diam [minvol])

Parameters:

node	node ID
elev	elevation of bottom of tank where water level is zero, ft (m)
initlevel	initial water level above tank bottom, ft (m)
minlevel	lowest allowable water level, ft (m)
maxlevel	highest allowable water level, ft (m)
diam	tank diameter, ft. (m)
minlevel	volume of water below minimum level, cu ft (cu m)

Remarks:
One line should appear for each tank or reservoir.

For reservoirs you need only enter the node ID and elevation. By definition, the water surface in a reservoir remains at a fixed elevation while it varies in a tank as flows enter or leave.

Water surface elevation in tanks equals the bottom elevation plus water level. Tanks are assumed to be cylindrical between their minimum and maximum levels. Non-cylindrical bottom sections can be accommodated by supplying the volume of the section as the last parameter on the line. For non-circular tanks, use a diameter equal to 1.128 times the square root of the area.

A [TANKS] section is required.

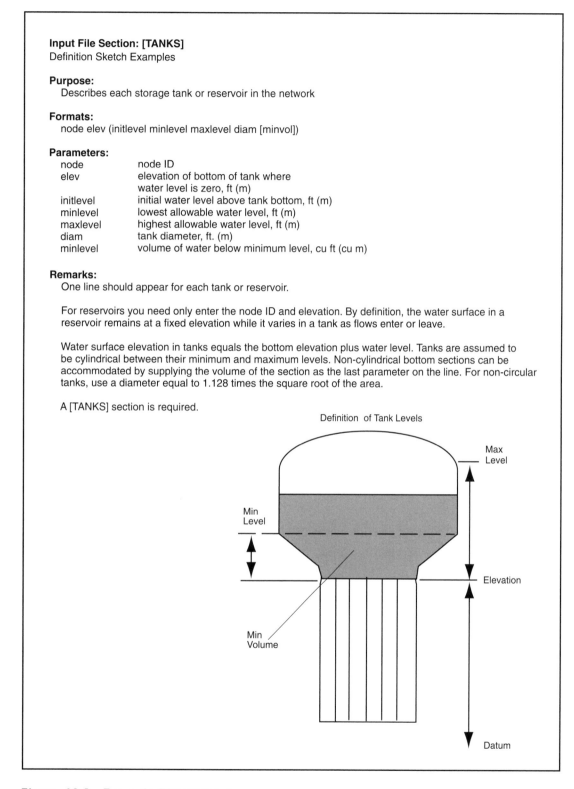

Definition of Tank Levels

Figure 10-3 Example EPANET help screen

Flexibility

There is often some desired feature not currently supported by a model. In most cases, the user is totally constrained by the capabilities of the existing packages. In some packages, however, there is some flexibility to modify or expand the analysis and reporting capabilities. Complete flexibility to modify the model is only possible if the model source code is available (generally the case only for public domain models).

Training

Training materials for models may include books, manuals, videos, and hands-on training courses. For many software packages, open training courses are offered by the model distributor. Many of these distributors will also provide special classes either at the distributors' offices or in the office of the client. In most cases, these courses provide some general background on modeling, but usually emphasize the specific hands-on training for a specific model.

MODEL SELECTION STRATEGY

Two basic strategies typically used to select a model are (1) conduct a detailed, in-depth study and comparison of models (possibly accompanied by a call for proposals prior to selecting the model) or (2) review the available information and communicate with others who are engaged in the same process. Though the second method may seem nonscientific, under many circumstances it may be the most appropriate approach.

Review of Available Models

There are many water distribution system models available for performing hydraulic analysis. Many of these models have been specifically developed or modified to accommodate water quality modeling. Table 10-1 contains a list of water distribution system models available at the time of this writing that have water quality modeling capabilities. Some of these models allow for only rudimentary, steady state water quality or time-of-travel analysis. The water quality modeling field is growing and new models are continually being introduced.

NETWORK REPRESENTATION

As mentioned at the beginning of this chapter, methods and protocols for representing a distribution system network are important components of all modeling work. In representing a distribution system in any hydraulic network model, the following types of information are required:

- pipe lengths, diameters, and roughness coefficients

- node elevations

- tank and/or reservoir sizes (i.e., diameter or dimensions, and minimum and maximum water level elevations)

- pump characteristics

- major valve characteristics

- water consumption patterns

- tank, pump, and valve operational information

- source quantity patterns

Table 10-1   Water quality modeling programs (as of Dec. 2, 1997)

Model	Distributor	Internet/e-mail
AQUA	Computer Modeling, Inc., 2121 Front St., Cuyahoga Falls, OH 44221, (330) 929-7886	jwc@gwis.com
Cybernet	Haestad Methods, Inc., 37 Brookside Rd., Waterbury, CT 06708, (800) 727-6555	www.haestad.com
EPANET	Dr. Lewis Rossman, USEPA, 26 W. Martin Luther King Dr., Cincinnati, OH 45268, (513) 569-7603	www.epa.gov/ORD/NRMRL/ epanet/ rossman.lewis@epamail.epa.gov
FAAST-3	Faast Software, 3062 East Ave., Livermore, CA 94550, (510) 455-8086	www.faast.com
H2ONET	MW Soft, Inc., 300 N. Lake Ave., Suite 1200, Pasadena, CA 91101, (800) 442-0638	www.mwsoft.mw.com
KYPIPE3	Civil Engineering Software Center, 354 Civil Engineering Building, University of Kentucky, Lexington, KY 40506-0281, (606) 257-8005	www.engr.uky.edu/CE/KYPIPE
PICCOLO	Safege Consulting Engineers, 15/27 rue du Port, BP 727, F-92007 Nanterre Cedex, France, +33 1 46 14 71 83	piccolo@safege.attmail.com
Stoner Workstation Service	Stoner Associates, Inc., P.O. Box 86, Carlisle, PA 17013, (717) 243-1900	www.stoner.com
TDHNET	TDH Engineering, 607 Ninth St., Laurel, MD 20707, (301) 490-4515	timhirrel@compuserve.com
WaterCAD	Haestad Methods, Inc., 37 Brookside Rd., Waterbury, CT 06708, (800) 727-6555	www.haestad.com
WaterMax	The Pitometer Associates, 20 North Wacker Dr., Chicago, IL 60606, (800) 347-5990	
WaterWorks	Aquateq Research, 50-2450 Hawthorne Ave., Port Coquitlam, B.C., Canada V3C 6B3	dale@radiant.net
WATNET	WRc, 8 Neshaminy Interplex, Suite 219, Trevose, PA 19053, (215) 244-9972	www.wrcgroup.com

The following additional information that specifically effects distribution system operation are required:

- water quality patterns at sources

- dynamic behavior parameters of substances in the distribution system

Two sources of background information on representing a network in a hydraulic model include Cesario (1995) and AWWA (1989).

Development of the additional parameters and information used in water quality models is discussed in earlier chapters in this book.

Calibration

In the calibration step, various model parameters are adjusted so that the results of the model reflect observed field measurements. In some cases, the network representation may be modified in the calibration step. This step is required because there is uncertainty in the results of any model application due to approximations associated with the formulation of the model and unknowns in some model parameters. An example of uncertainty in the model application is the assumption of complete mixing at all nodes. An example of an approximation in parameter values is the use of nominal pipe diameters that may be in error due to tuberculation, which tends to reduce the actual diameter of pipes over time.

Calibration has been the subject of much discussion and many papers. Two recent papers on calibration are Cesario et al. (1996) and Walski (1995), listed in the references section.

Several methods are used for calibrating hydraulic and water quality models. These include both traditional hydraulic field methods and the more recently developed use of tracers in conjunction with hydraulic and water quality models. The traditional methods include extensive measurement of flows and pressures throughout the distribution system and/or measurement of C-factors for representative pipes.

A new method of calibration relies on tracer studies used in conjunction with water quality models. Water quality models are used to predict the movement and transformation of substances, but all distribution system water quality models depend on hydraulic models to provide information on the direction and magnitude of pipe flows. If there are inaccuracies in flow and/or velocity values provided by the hydraulic model, then the water quality predictions will also be inaccurate.

Though most water quality models can be used to represent both conservative and nonconservative substances, the use of conservative substances is most appropriate for calibration of hydraulic models. When modeling a conservative substance, there are essentially no water quality parameters to be adjusted. In other words, if the hydraulic parameters are correct and the initial conditions and loading conditions for the substance are known, then the water quality model should provide a good estimate of the concentration of a substance throughout the network.

The calibration process using water quality modeling may be summarized as follows:

1. A conservative tracer is identified for a distribution system. This tracer may be one that is specifically added to the distribution system, such as fluoride, or when there are multiple sources of water, a naturally occurring substance in the water source, such as hardness.

2. A controlled field experiment is performed in which (a) the conservative tracer is injected into the system for a prescribed period of time; (b) a conservative substance that is normally added, such as fluoride, is shut off for a prescribed period; or (c) a naturally occurring substance in one of the sources is traced.

3. During the field experiment, the concentration of the tracer is measured at selected locations in the distribution system along with other operational parameters required by a hydraulic model, such as tank water levels, pump operations, flows, etc. In addition to the conservative tracer, other water quality concentrations, such as chlorine residual, may be measured, although these values are not generally used in the calibration process.

4. Standard means are used to adjust the parameters in the hydraulic model to represent the operations of the distribution system. The water quality model is then used to model the conservative tracer.

5. If the model adequately represents the observed concentrations, then it is likely that the hydraulic model is well calibrated for the conditions being studied. If there is significant deviation between the observed and modeled concentrations, then further calibration of the hydraulic model is required. Various statistical and directed search techniques have been used in conjunction with conservative tracer data to aid the user in adjusting the hydraulic model parameters in order to match the observed concentrations.

The tracer–water quality modeling technique has been employed in several studies around the world to assist in calibrating hydraulic models. Several of these studies were discussed earlier in this book. Though conservative tracer–water quality modeling has not been as widely used as the more classical calibration methods, the studies that have been performed have demonstrated the feasibility of this approach for calibrating hydraulic models.

Accessing EPANET

Most of the discussion in this book has focused on the use of EPANET, which is a public sector water quality model. EPANET is a computer program that performs extended period simulation (EPS) of hydraulic and water quality behavior within drinking water distribution systems. It tracks the flow of water in each pipe, the pressure at each pipe junction, the height of water in each storage tank, and the concentration of a substance throughout a distribution system during a multitime period simulation. In addition to substance concentrations, water age and source tracing can also be performed. The water quality module of EPANET is equipped to model such phenomena as reactions within the bulk flow, reactions at the pipe wall, and mass transport between the bulk flow and the pipe wall. Complete documentation on the use of EPANET on a personal computer under DOS and Microsoft Windows® and the program itself is available through the Internet USEPA home page. The Internet address is

http://www.epa.gov/epahome/datatool.htm#water

Under Windows, the user is able to edit EPANET input files, run a simulation, display the results on a color-coded map of the distribution system, and generate additional tubular and graphical views of these results.

References

AWWA. 1989. *AWWA Manual M32, Distribution Network Analysis for Water Utilities.* Denver, Colo.: American Water Works Association.

Cesario, L. 1995. *Modeling, Analysis, and Design of Water Distribution Systems.* Denver, Colo.: American Water Works Association.

Cesario, L., J.R. Kroon, W.M. Grayman, and G. Wright. 1996. New Perspectives on Calibration of Treated Water Distribution Systems. In *Proc. AWWA 1996 Annual Conference.* Denver, Colo.: American Water Works Association.

Walski, T.M. 1995. Standards for Model Calibration. In *Proc. 1995 AWWA Computer Conference.* Denver, Colo.: American Water Works Association.

A

Equations for One-, Two-, and Three-Compartment Models

One-Compartment Model

Inflow Conditions: Water is assumed to enter the tank at a flow rate of Q_{in} and fluoride concentration of C_{in} (Clark et al. 1996). The value of C_T is the concentration of fluoride in the tank and V_T is the volume of water in the tank. Therefore, the net concentration of fluoride in the tank during the inflow period would be $(C_{in} - C_T)$. There is no outflow during this stage. The initial conditions in the tank are $C_T(t = 0) = C_{T_0}$ and $V_T(t = 0) = V_{T_0}$. By substituting these values in Eq 9-1 and 9-2 in chapter 9 we obtain

$$\frac{dC_T}{dt} = \frac{Q_{in}(C_{in} - C_T)}{V_T} \qquad \text{(Eq A-1)}$$

Outflow Conditions: Water flows out of the tank at a rate of $-Q_{out}$ and there is no inflow to the tank. Therefore, by substituting the appropriate values for the flow and fluoride concentration in Eq 9-1 and 9-2 we will have

$$\frac{dC_T}{dt} = 0 \qquad \text{(Eq A-2)}$$

Two-Compartment Model

Inflow Conditions: The volume of water in compartment A, V_A, is fixed. Compartment B has a variable water volume, V_B. Water enters compartment A with a flow of Q_{in} gallons per minute and with a fluoride concentration of C_{in}. Flow rate from compartment A to B is indicated by Q_{AB}, and C_B indicates the fluoride concentration in tank B. The initial conditions for compartment A are $C_A(t=0) = C_{A_0}$ and $V_A(t=0) = V_{A_0}$. Since compartment A has a fixed water volume, then $\dfrac{dV_A}{dt} = 0$ and $Q_{in} = Q_{AB}$. By substituting these conditions in Eq A-1 and A-2, we will have

$$\frac{dC_A}{dt} = \frac{Q_{in}(C_{in} - C_A)}{V_A} \qquad \text{(Eq A-3)}$$

Inflow conditions for compartment B with $\dfrac{dV_B}{dt} = Q_{AB}C_A$ result in

$$\frac{dC_B}{dt} = \frac{Q_{in}(C_A - C_B)}{V_B^{(n)}} \qquad \text{(Eq A-4)}$$

Outflow Conditions: Water flows out of the compartment B at a rate of $Q_{AB} = Q_{out}$ and a fluoride concentration of C_B. By substituting these values into Eq 9-1 and 9-2, we have

$$\frac{dC_B}{dt} = \frac{Q_{AB}(C_B - C_B)}{V_B^{(n)}} = 0 \qquad \text{(Eq A-5)}$$

The mass balance equations for outflow from compartment A, with an outflow rate of $Q_{out} = Q_{BA}$, will imply

$$\frac{dC_A}{dt} = \frac{Q_{out}(C_B - C_A)}{V_A} \qquad \text{(Eq A-6)}$$

Three-Compartment Model

For this model it is assumed that the volumes of compartments A and C are fixed, and the volume of compartment B is variable and a function of time t, and there is a flow exchange between compartments B and C, represented by Q_{BC} and Q_{CB}. This flow rate can be adjusted to a fraction of the inflow or outflow, depending on the condition of the tank at a given time.

Inflow Conditions: Since compartments A and C have fixed volumes, then $Q_{in} = Q_{AB}$, $Q_{BC} = Q_C$, and $Q_{BA} = Q_{out}$. The mass balance for inflow for compartment A will result in

$$\frac{dC_A}{dt} = \frac{Q_{in}(C_{in} - C_A)}{V_A} \qquad \text{(Eq A-7)}$$

By substituting the net flow values for compartments B and C into Eq 9-1 and 9-2 we obtain

$$\frac{dC_B}{dt} = \frac{Q_{in}(C_A - C_B) + Q_{BC}(C_C - C_B)}{V_B^{(n)}}$$ (Eq A-8)

$$\frac{dC_C}{dt} = \frac{Q_{BC}(C_B - C_C)}{V_C}$$ (Eq A-9)

Outflow Conditions: By substituting the outflow conditions for compartment C, with a fixed water volume, and compartments B and A, we have the following:

$$\frac{dC_C}{dt} = \frac{Q_{BC}(C_B - C_C)}{V_C}$$ (Eq A-10)

$$\frac{dC_B}{dt} = \frac{Q_{BC}(C_C - C_B)}{V_B^{(n)}}$$ (Eq A-11)

$$\frac{dC_A}{dt} = \frac{Q_{out}(C_B - C_A)}{V_A}$$ (Eq A-12)

FLOW-THROUGH TANK MODEL

As with the classical tank model, these compartments were used to represent the mixing regime in the flow-through tank model. Figure A-1 illustrates the flow conditions associated with these assumptions.

One-Compartment Model

In this case the tank is filling and emptying at the same time. Therefore, the change in the volume is a function of the difference between inflow and outflow, i.e., $\frac{dV}{dt} = Q_{in} - Q_{out}$. After appropriate adjustments in Eq 9-1 and 9-2, we have

$$\frac{dC_A}{dt} = \frac{Q_{in}(C_{in} - C_A)}{V_A^{(n)}}$$ (Eq A-13)

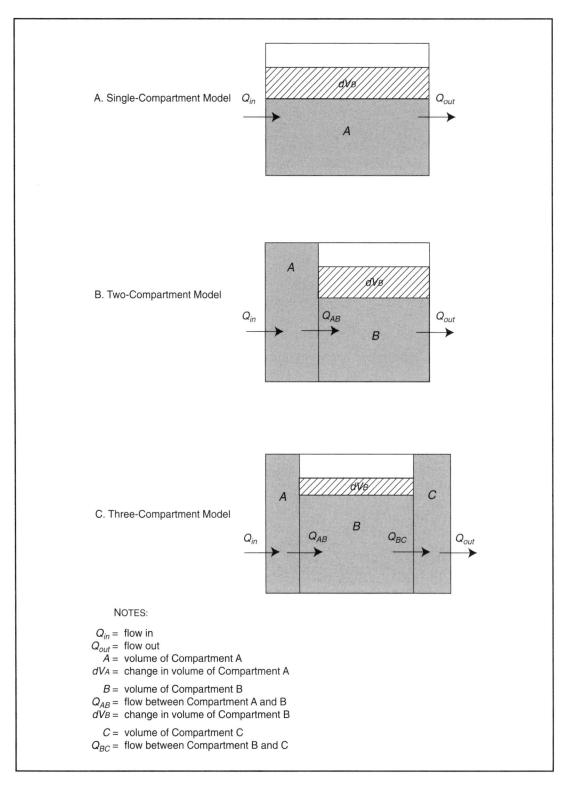

A. Single-Compartment Model

B. Two-Compartment Model

C. Three-Compartment Model

NOTES:

Q_{in} = flow in
Q_{out} = flow out
A = volume of Compartment A
dV_A = change in volume of Compartment A

B = volume of Compartment B
Q_{AB} = flow between Compartment A and B
dV_B = change in volume of Compartment B

C = volume of Compartment C
Q_{BC} = flow between Compartment B and C

Figure A-1   Flow-through tank models

Two-Compartment Model

In this model, the volume of compartment A is fixed, therefore $\dfrac{dV_A}{dt} = 0$ and $Q_{in} = Q_{AB}$. Then for compartment A we have

$$\frac{dC_A}{dt} = \frac{Q_{in}(C_{in} - C_A)}{V_A} \qquad \text{(Eq A-14)}$$

The volume in compartment B is variable and is a function of inflow Q_{in} and outflow Q_{out}. Therefore, the change of fluoride concentration is $Q_{in}C_A - Q_{out}C_B$. Substituting these values in Eq 9-1 and 9-2 will result in

$$\frac{dC_B}{dt} = \frac{Q_{in}(C_A - C_B)}{V_B^{(n)}} \qquad \text{(Eq A-15)}$$

Three-Compartment Model

The model assumes constant and equal volumes for compartments A and C. Therefore, the inflow to compartment A is the same as the flow from compartment A to B, $Q_{in} = Q_{AB}$. For the same reason, flow from B to C is the same as the outflow from compartment C, $Q_{BC} = Q_{CB}$. Compartment B has a variable volume that is a function of the difference between inflow and outflow. Considering these conditions for compartments A, B, and C we will have

$$\frac{dC_A}{dt} = \frac{Q_{in}(C_{in} - C_A)}{V_A} \qquad \text{(Eq A-16)}$$

$$\frac{dC_B}{dt} = \frac{Q_{in}(C_A - C_B)}{V_B^{(n)}} \qquad \text{(Eq A-17)}$$

$$\frac{dC_C}{dt} = \frac{Q_{out}(C_B - C_C)}{V_C} \qquad \text{(Eq A-18)}$$

Reference

Clark, R.M., F. Abdesaken, P.F. Boulos, and R. Mau. 1996. Mixing in Distribution System Storage Tanks: Its Effect on Water Quality. *Jour. Environ. Eng.*, 122(9):814–821.

Glossary

Bulk water phase The water flowing through a pipe.

CAD Computer-assisted design.

Conservative A constituent that does not degrade.

Dynamic modeling A form of modeling in which demand and supply are allowed to vary with time and the resulting solution is determined.

Event-driven method A water quality model based on a next-event scheduling approach.

Extended period simulation (EPS) Hydraulic and/or water quality simulation in which the time periods for simulation are broken down into small increments or periods in order to approximate continuous variation in distribution demand conditions.

First-order decay A condition described by a differential equation in which the rate of decay of a constituent is a function of the concentration of the constituent to the first power.

GIS Geographic information systems.

GL Gigalitres or a billion litres.

GUI Graphical user interface.

Hydraulic balancing The technique of solving a set of equations that describe the head loss across a given set of links in a distribution model, in which the head loss is described as a function of flow in the link.

Hydraulic network models Models developed to simulate flow and pressures in a distribution system, either under steady state conditions or under time-varying conditions.

***k* rates** A decay coefficient in a first-order equation.

Limited first-order decay A condition described by a differential equation in which the rate of decay of a constituent is a function of the concentration of the constituent to the first power minus a constant value.

Link The graphical representation of a run of distribution system pipe between intersections.

Link–node network The graphical representation of a distribution system in which the pipes are represented as links and the junctions of pipes, wells, tanks, or starts of pipes as nodes.

Mass transfer concepts Transfer of one substance through another on a molecular scale.

Model calibration The process of adjusting the model input data (or, in some cases, model structure) so that the simulated hydraulic and water quality outputs from the model mirror observed field observations.

Multiple-barrier concept A phrase used in water treatment to refer to more than one process or series of processes used to protect drinking water against contamination. For example, conventional treatment or filtration in combination with disinfection are referred to as multiple barriers.

***nth*-order decay** A condition described by a differential equation in which the rate of decay of a constituent is a function of the concentration of the constituent to the *nth* power.

Node A graphical representation of the intersection of two or more pipes in a distribution system.

Optimization model A mathematical means of generating a large number of solutions and selecting the solution that best fills an objective within specified constraints.

Parallel first-order decay A condition described by a differential equation in which the rate of decay of a constituent is a function of the concentration of two independently decaying constituents to the first power.

Perfect mixing A condition in which a tracer introduced into a container of liquid is assumed to be equally and instantaneously mixed throughout the container.

Pipe-wall demand The demand for chlorine or another disinfectant due to the interaction of the disinfectant itself with material on the pipe wall or with the pipe wall material itself.

Piston flow A condition in which no mixing occurs in the flowing liquid; frequently referred to as a plug flow.

SCADA Supervisory control and data acquisition system.

Skeletonization A representation of a distribution network showing only the major hydraulic elements of the network, such as major transmission lines, tanks, demands, and supplies. It is intended to represent the general behavior of the network.

Steady state modeling A form of modeling in which external forces are constant over time and yields solutions that would occur if the system were allowed to reach equilibrium.

Temporal variation Variation over time.

Temporally varying dynamic models Models that allow for variation over time.

Time-varying solution In dynamic modeling, demands and supplies are allowed to vary over time. The solution therefore varies over time.

Water quality tracers Naturally occurring or deliberately added chemical tracers measured in the field. The results are used to calibrate hydraulic and water quality models.

Index

An *f.* indicates a figure, a *t.* indicates a table.